I0048852

TELEROBOTIC MARS EXPEDITION DESIGN

New Ways to Explore Mars

Edited by
Dr. Frank Crossman
The Mars Society

Published by Polaris Books
Lakewood, Colorado

THE MARS SOCIETY

The chapters of this book are copyright © 2022 by the authors of the chapters. The authors have licensed The Mars Society to publish this work directly or through a commercial publisher and to use it for all of The Mars Society's current and future print and electronic uses.

This softcover edition is Copyright © 2023 by The Mars Society

ISBN - 978-1-7363860-4-0

TABLE OF CONTENTS

PREFACE   v
     Robert Zubrin

1: A MIFECO HEAVY LIFT MARS ROBOTIC MISSION   1
     Robert Mills

2: PROJECT LATE (Lava Tubes Exploration)   25
     Piotr Torchała, Justyna Pelc, Beata Suścicka, Ewa Borowska,
     Magdalena Łabowska, Nikola Bukowiecka, Aleksandra Klassa,
     Bartosz Rybacki, Hubert Gross, Nikodem Drąg, Michał
     Słomiany, Kamil Ziółkowski, Kajetan Szostek,
     Milena Michalska

3: PRECURSOR - The Ultimate Telerobotic Precursor to a Human   45
Mission to Mars: Rollers, Rotors & Unfinished Business
     Carl Greenbaum

4: TELEROBOTIC MARS MISSION FOR LAVA TUBE EXPLOR-   67
ATION AND EXAMINATION OF LIFE
     Hanjo Schnellbaecher, Florian Dufresne, Tommy Nilsson,
     Leonie Becker, Oliver Bensch, Enrico Guerra, Wafa Sadri,
     Vanessa Neumann

5: THE VULCAN FORGE MISSION   93
     Alden R. O'Cain, Neel S. Shah, Jason Thai· Isaac W.R. Bahler·
     Kai A. Fucile Ladouceur, Samantha B. Chong

6: FOUNDATIONS FOR HUMANS ON MARS   119
     Mark K. Moran, Anjali Saini

PREFACE

Robert Zubrin

Over the past three decades, NASA has conducted a series of highly successful robotic exploration missions to the Red Planet. These have consisted of orbiters or rovers (and most recently a subscale experimental helicopter) launched singly or in pairs every two to four years, with a total of three static landers (*Pathfinder, Phoenix, and Insight*) and five rovers of increasing capability (*Sojourner, Spirit, Opportunity, Curiosity, and Perseverance*) reaching Mars over the period in question.

We have learned a lot from this epic program, but the time has come to do more. While robotic rovers are wonderful, they cannot resolve the fundamental scientific questions that Mars poses to humanity, which relate to the potential prevalence and diversity of life in the universe. We now know that the early Mars was very much like the early Earth; a rocky warm and wet planet with a carbon dioxide dominated atmosphere.

That raises critical questions. Life appeared on Earth virtually as soon as our planet was cool enough for liquid water. So did it appear on Mars, too? If so, did it use the same DNA-RNA information system underlying all life on Earth, or something else?

NASA Kepler Space Telescope has revealed that some twenty percent of the 400 billion stars in our galaxy also have Earth sixed planets in their habitable zone. Is life likely to be found everywhere? Is life as we know it on Earth what life is? Or is it just a particular example drawn from a vast tapestry of possibilities?

These are questions that thinking men and women have wondered about for thousands of years. They can only be resolved by sending humans to Mars.

Finding evidence of past life requires fossil hunting. *Perseverance* will make a stab at that, but human rockhounds, capable of traveling far over difficult terrain, climbing, digging, doing delicate work, and intuitively following up clues can do that job vastly better. Finding extant life to determine its nature will require drilling down hundreds of meters to reach underground water where life might still thrive, bringing up samples, culturing them, and subjecting them to analysis. That is light years beyond the ability of robotic rovers.

If we don't go, we won't know.

We need to start developing the hardware that can take humans to Mars.

Sending humans to Mars does not require building gigantic multi megawatt ion drive science fiction spaceships in a futuristic world of orbital spaceports. It requires sending a payload of 10 tons or more capable of supporting a small group of people from Earth to Mars, landing it, and then sending that or a comparable payload back.

The currently operational SpaceX Falcon Heavy could throw a 10-ton class lander to Mars. Blue Origin's New Glenn should be able to do about the same. Even more powerful launch vehicles will soon be operational. As I write these lines, NASA's SLS has just demonstrated its first successful test flight, sending an Orion capsule on a voyage around the Moon, and the SpaceX Starship booster is being prepared for its first flight test. Both of these will be able to send a 20-ton lander. So we have that part of the transportation system covered.

The key missing piece is the lander.

The Perseverance landing system can deliver one ton to the surface of Mars. To get started with human exploration, we need a 10-ton class lander. There are a number of ways to create such a system. For example, we could use aeroshells, parachutes and landing jets, or perhaps a miniature version of Starship. I won't go into the details. But the bottom line is if we can land one ton on Mars we can land 10. It requires no scientific breakthroughs. But it does need to be developed.

There are other key hardware elements needed before we send humans, but the combination of heavy lift with a heavy payload lander can get us started. Once we have a 10-ton lander, we can use it to send large robotic expeditions to Mars. Instead of landing one rover, we will be able to land platoons of robots.

The possibilities this opens are enormous. For example, a robotic Mars expedition could include science explorers like *Perseverance*, and much bigger versions of the *Ingenuity* helicopter capable of wide-ranging reconnaissance. Smaller rovers armed with high-resolution cameras could create a high-definition map of the area, transmit to Earth, to allow millions of people here walk the landscape with virtual reality gear, directly assisting the robots in exploration by calling their attention to features of interest.

The expedition could also include construction robots, possibly humanoid in form with arms and legs, capable of building a Mars base. These could set up a power system and put in operation units for converting Martian carbon dioxide and water ice into methane and oxygen rocket propellant and store it in tanks. With such a base set up and fully equipped with housing, power, a well-instrumented lab, a workshop, and supplies in advance, all astronauts will need to do is show up in their well-proven lander armed with a credit card, and check in. Everything they need to live and work on Mars, and return from Mars, will be there waiting for them.

By taking this approach we can tackle the central challenges facing human Mars missions head on while greatly enhancing our scientific exploration capability immediately. So that is what we should do.

But as noted, there are many options for the design of such missions. These need to be explored in order to get the best possible result. Therefore, in order to facilitate such an intellectual exchange, in 2022 the Mars Society sponsored a contest for the design of a telerobotic Mars exploration expedition. Contestants were not asked to design the flight system, as all too often advanced Mars missions designers focus on that to the exclusion of what the mission actually accomplishes on the Martian surface. (NASA's current Design Reference Mission for human exploration being a particularly egregious example of that, spending two years in transit and only 30 days on Mars.) Rather, contestants were told that they had lander capable of delivering 10 tons of gear to the Martian surface and take it from there, with points being awarded for the designs merit in the areas of engineering soundness, science return, contribution to preparing future human missions, and cost. Seventeen teams submitted designs. These were downselected to six finalists who presented their designs to a panel of distinguished judges at the International Mars Society Convention at Arizona State University in October 2022. All six of those finalist designs are presented here.

The first-place design, by veteran American engineer Robert Mills, is strongly focused on the near-term possibilities, as it looks at combined operations between types of robots that are currently in use. Very importantly, Mills refutes the contention that the cost of a mission involving large numbers of robots would be astronomical. On the contrary, he shows how by taking advantage of the economies of multiple unit production and using redundancy to allow acceptance of risk of failure to any one robot, costs can be kept well under control, even as science return is multiplied exponentially.

The second-place winner, by a team of Polish engineers is more visionary, but extremely interesting, as it presents an exciting idea of the amazing sorts things the future could bring as the robotic capabilities evolve.

It was hard to beat two such strong competitors. But each of the remaining semi-finalists advanced ideas that are well worth considering as part of the mix.

So where do we go from here? I think NASA should build on these ideas and take the next step by holding an official workshop to collect concepts from all corners of the industry for the design of robotic Mars expeditions taking advantage of the capabilities of a ten-ton payload class Mars lander. Using the best of those ideas, the agency should then develop a design concept for a ten-ton payload robotic Mars expedition. That design would provide the requirements to develop the lander we need.

Then start flying robotic expeditions that really open the way to Mars.

Golden, Colorado February 27, 2023

1: A MIFECO HEAVY LIFT MARS ROBOTIC MISSION

Robert Mills
MIFECO Executive Consulting
Grand Haven, MI
robertstar2000@gmail.com

ABSTRACT

This mission design features the use of VR and digital twinning to allow the teleoperation of multiple devices while mitigating bandwidth and Mars to Earth transmission delays. This facilitates multipurpose rovers operating at higher speeds than current rovers able to perform general purpose exploration and construction, it also allows the operation of a general purpose robotic laboratory capability located on Mars. The MIFECO company is a boutique futurist engineering consulting firm specializing in technological strategy with deep expertise in engineering and information science.

INTRODUCTION

The Proposed Mission

An aggressive mission schedule is laid out with well-constrained risk, addressing both significant science objectives and leaving in place infrastructure that will be needed for future missions. The Mars society framework indicates planning for a very aggressive 5-year development providing additional time to get started and to cover unforeseen issues. Doing a worthwhile program in this timeframe precludes developing many next-generation technologies, so the proposed mission will use existing hardware technology enhanced with innovations in operations and software to provide mission capabilities far beyond those that have been seen to date. The development of this program will follow the SpaceX™ development strategy - doing very rapid development pragmatically while allowing opportunities for problems to drive iterative improvements during an Earth-based testing phase. This approach will provide a solution that has a high probability of meeting the program requirements and timeline, even considering unforeseen program issues.

The integration with a launch vehicle and the delivery of the robotics to Mars is not covered.

Expedition Objectives

The expedition objectives are summarized as:

• Explore, find and map out scientifically relevant geological features, biological markers, and areas of future interest over a large area. The proposed mission will

use macro and microscopic imaging, surface and subsurface mapping using LIDAR, radio-based navigation, ground-penetrating radar, acoustic subsurface imaging using small explosives, multiple sensors, and sample taking (with analysis of samples included in this plan)
- Find and retrieve usable amounts of water, analyze it for impurities, while piloting filtration, storage, and electrolysis.
- Acquire the data needed to do a deep and accurate scientific study of new and exciting finds. Provide the ability to emulate robotically many of the processes commonly used by geologists, biologists, and other disciplines. Include the ability to evolve the experiments as the scientists explore.
- Be able to return to interesting locations for deeper analysis and better samples
- Provide a mobile Mars base for recharging rovers and devices, storing equipment, running various laboratory experiments and analyses, and to provide a future habitat and garage for future unmanned and crewed expeditions.
- Perform lab analysis of minerals, samples of water, air, while looking for signs of life (utilizing the Mars based laboratory capability).
- Collect and store samples for future return missions.
- Identify locations for future missions and build future supporting infrastructure (including finding and leveling landing fields, providing landing navigation beacons, finding a local sources of water, electrolyzing it into stored oxygen, providing a long life - Mars-based power source, providing robotic capabilities to perform future construction, and leaving in place future habitats and/or a future equipment storage garages.
- All the data collected would (of course) be transmitted to Earth for additional processing and analysis.
- Complete the mission objectives within a 1-2-year timeframe on Mars preparing for the next round of missions.

THE PLAN

The Mission Plan is to provide a highly general-purpose partially teleoperated and partially autonomous robotic capability that can perform a wide variety of pre-defined and ad-hoc mission objectives. To accomplish this the robot rovers will use several innovations to vastly increase the breadth of capabilities and speed compared to current Mars robotic designs. These include a method to allow teleoperation with minimal impact from communication delays as well as several other innovations. This expedition will use a group of eight identical ruggedized robotic rovers and two mobile stations plus various tools.

Using Virtual Reality to navigate through a digital twin of the Martian surface is a breakthrough concept providing both data compression & consolidation of data from multiple devices and rovers, this will allow simultaneous device control and accommodate bandwidth & the significant lightspeed communications delay

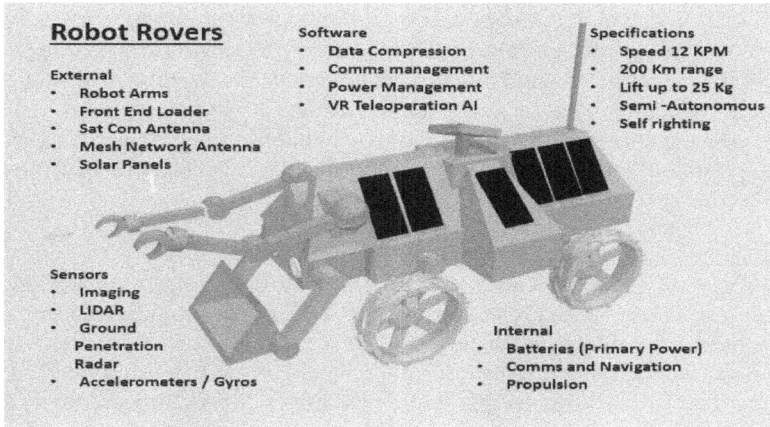

Robot Rovers

External
- Robot Arms
- Front End Loader
- Sat Com Antenna
- Mesh Network Antenna
- Solar Panels

Software
- Data Compression
- Comms management
- Power Management
- VR Teleoperation AI

Specifications
- Speed 12 KPM
- 200 Km range
- Lift up to 25 Kg
- Semi -Autonomous
- Self righting

Sensors
- Imaging
- LIDAR
- Ground Penetration Radar
- Accelerometers / Gyros

Internal
- Batteries (Primary Power)
- Comms and Navigation
- Propulsion

Figure 1 Robotic Rover

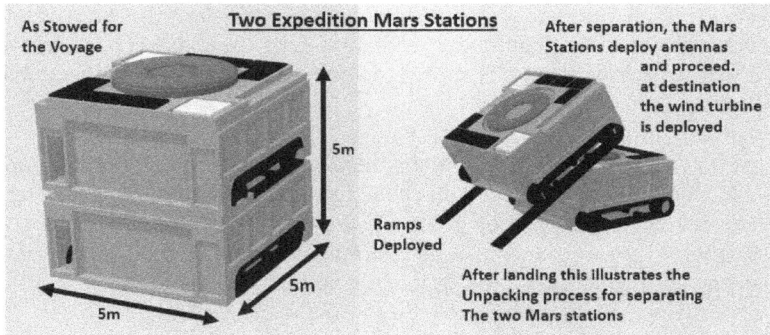

Two Expedition Mars Stations

As Stowed for the Voyage

After separation, the Mars Stations deploy antennas and proceed. at destination the wind turbine is deployed

5m

Ramps Deployed

5m

5m

After landing this illustrates the Unpacking process for separating The two Mars stations

Figure 2 - Mars Station stacked for flight

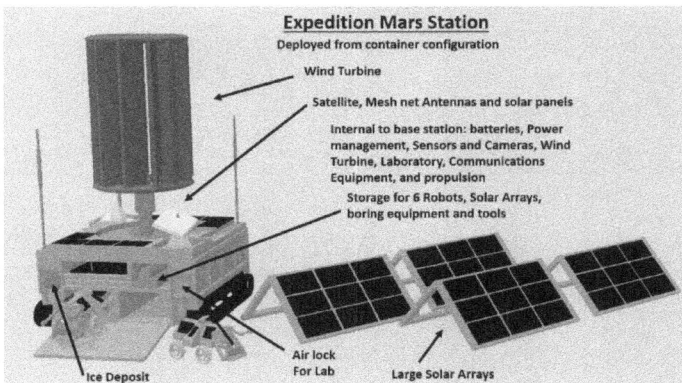

Expedition Mars Station
Deployed from container configuration

Wind Turbine

Satellite, Mesh net Antennas and solar panels

Internal to base station: batteries, Power management, Sensors and Cameras, Wind Turbine, Laboratory, Communications Equipment, and propulsion

Storage for 6 Robots, Solar Arrays, boring equipment and tools

Ice Deposit

Air lock For Lab

Large Solar Arrays

Figure 3 - Mars Station, Rover & Solar Arrays

The mission includes two mobile transport containers / used on Mars as mobile Mars stations. They include storage for: robotic rovers, solar array modules, tools, and include an integrated wind turbine. Two methods of cleaning the dust that will inevitably collect on the solar arrays are described later in this proposal The Mars station also contain much of the mission computing, power storage, a robotic laboratory, and Mars to Earth communications. The Mars stations have mobility capabilities.

<u>Robotic Rovers</u> are designed to operate at a higher speed and with more fluid operator control than any current rovers and will do so *using Virtual Reality navigating in a digital twin of the local Martin environment* providing teleoperation that seems like real time to teleoperators. This will allow teleoperators to drive and manipulate objects in a way that simulates normal Earth-based remote teleoperation. This will reduce the effect of Earth to Mars communication delay and bandwidth limitations allowing multiple operators teleoperating multiple rovers to accomplish vastly more scientific work than currently imagined possible. The rovers include LiDAR to map the Mars environment to create the virtual environment. Software will compress the LiDAR data by providing only data representing changes. High and Low-resolution imaging is avalible and combined with LiDAR to provide more VR realism. High-resolution photos will avalible for scientific study.

<u>Proposed Innovations</u> Through explaining the features of our robotics and how they utilize unique innovations it is intended to convince the reader that it is possible to achieve the expedition goals. The design has intentionally avoided the use of technology not currently available or that has unpredictable development times such as fully autonomous AI or nuclear power.

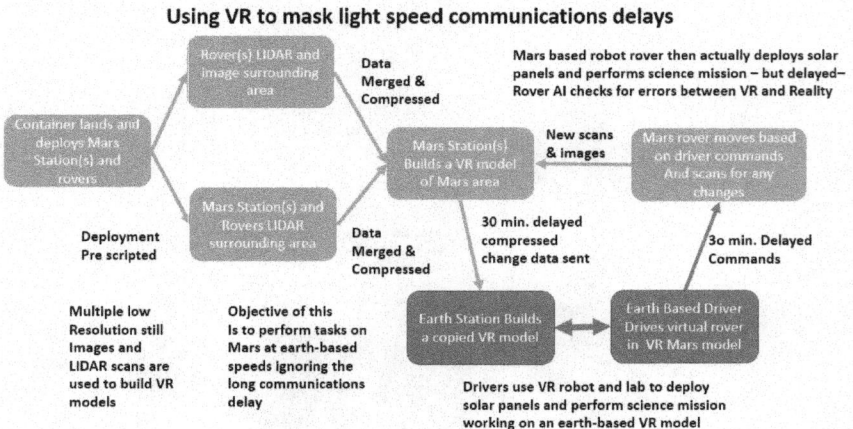

Figure 4 Using Digital Twinning and Virtual Reality to overcome delay

Addressing Remote Teleoperations using twinning and VR uses a new method to mitigate the Mars to Earth light-speed communication delay. This new method is a critical differentiation of this proposal allowing a dramatic improvement in capabilities. Combining sensor data and movement commands, a virtual model (a digital twin of the Martin environment, rovers, and devices is constructed on the Mars station and sent to Earth). Since the only things moving are those commanded to move from the Earth the model can be updated in real-time on earth as rovers move. Whenever any device is instructed to move the VR model is updated. As the rovers move, new compressed sensor data (sending only data not sent previously) will be sent to Earth to update and extend the virtual model. Teleoperators, seeing everything move in apparent real-time within the virtual model, will drive and use robotic arms in the virtual reality model as though they were on Mars using VR goggles and haptic feedback VR gloves. This same process also provides control of the Mars Station movement and the Mars based laboratory. Teleoperation of the laboratory and its instruments is modeled in VR in the same way as the rovers, allowing the same benefits and allowing scientists to remotely work in the laboratory much as they would on Earth setting up and operating analysis and experiments (then waiting for the results).

All movement commands will then be sent to the robotic devices on Mars. After the lightspeed communications lightspeed delay the commands will be executed. (The VR model will use the same physical rules (such as gravity) as the actual devices on Mars. As a safety net the on Mars AI will look for discrepancies between the expected results of the commands and reality. Teleoperators can drive and manipulate objects in virtual real-time in the virtual space allowing research to proceed rapidly thus, masking the lightspeed communications transit time (~7-40-minute light speed delay). Only a small amount of compressed data is needed to keep the digital twin aligned. with reality.

Using VR to enable teleoperations with a substantial communications delay inspired by an MIT team CSAIL who created Baxter's Homunculus: Virtual Reality Spaces for teleoperation in manufacturing. Associate Jeffrey Lipton et al arXiv

Long duration loss of communications, like solar occlusions, will still be an issue. A trained machine learning AI program will monitor any differences between the digital twin and reality to provide halt commands (Such as digging and hitting a large rock). In these cases, the operator will need to wait for the digital twin to re-synchronize with reality. Many functions (Such as: connecting robot rovers for recharge, cleaning solar panels, or retracing a path in the event of lost communications) will be fully automatous.

A Mars local mesh radio network is included in, the Robotic rovers, Mars stations, and in deployable repeaters, this will be used to increase local bandwidth and reduce power requirements. The mars stations and robotic rovers will also have satellite communications (as a backup) The mesh network will allow high bandwidth communications on the Mars surface, better reliability, and robustness, and reduce the power consumption and computing power needed for most communications. The two Mars stations (with access to more power and more powerful computing) will primarily support Earth to Mars communications.

Modular solar arrays and wind turbines are used as power sources. A common problem on Mars is that dust storms reduce the use of solar panels, sometimes for long periods. Using solar in conjunction with wind turbines and larger battery packs improves this as dust storms are usually accompanied by wind. Larger solar arrays are needed for power (compensating for 50% less than Earth efficiency). Arrays will be anchored to the ground by the robot rover using smart pitons (explained later) and will be connected to the Mars stations by cables put in place by the rovers. When needed individual arrays can recharge individual devices as each array will have its own power management electronics Each panel will have an electrostatic dust cleaning process[5] also a backup cleaning process using brushes via the robotic arms on rovers is also available. Using both processes combined with wind turbines will support long-duration operations.

Charging Rovers. On Mars recharging of rovers and other devices is another unique capability allowing rovers that can operate at greater power and speed (Up to 15 Kilometers per hour). Because they are not dependent on self-contained power sources they are able to go fast and lift large weights for construction. All Mars stations, rovers and tools have internal batteries and the ability to be recharged.

Mobile Mars Stations and Future habitat. Two multi-purpose mobile Mars stations would serve a variety of needs. They provide storage in-flight for all the rovers and devices, providing standby power and diagnostics during the voyage. During the initial deployment, they would be able to be relocated away from the landing vehicle and move to locations chosen for maximum scientific return using self-propulsion. Once positioned, they would deploy ramps and deploy rovers, the solar arrays and tools that would be used as needed during the mission. The stations would deploy a wind turbine and various antennas. They would also contain communications equipment, large capacity battery storage, onboard material science, and biology laboratory, hookups and cabling for the solar arrays, robotic rover charging station and the pilot water and oxygen processing equipment (including an aerogel insulated O^2 and H^2O tanks and equipment to melt and filter local ice). Further, into the mission, the Mars station would be buried in local soil and its internal shelving would be repurposed turning each Mars station into a future habitat (or garage) for support of future missions. These habitats would provide power, oxygen, process local water, and would include Mars to Earth communications and onboard laboratory capabilities.

Mission Objectives

Using the well-published needs of the scientific and aerospace community a list of potential objectives has been made. This list, constrained by the requirements of this expedition, and how rapidly technical capabilities can be created to support these objectives., Primarily created with existing technology. This all must be able to be developed and ready to fly in the required time frame. Below is a summary of the objectives that can, with an aggressive program, meet these criteria. Objectives for this proposal, including science and future infrastructure, listed below.

- Explore where targets of interest are and determine any that need to be explored in more detail or merit further study, acquire data beyond what existing rovers and satellite mapping have provided. (Use LADAR, imaging and using robotic arms to collect samples, using a small pickaxe, and using a boring rig (described latter in this chapter) to drill holes up to 4.5 meters deep). Perform subsurface mapping using small explosive charges (described with the smart pitons)
- Using rovers with ground-penetrating radar mapping, Examining near-surface lava tubes and caves to Map of sub-surface composition and density
- Obtain surface and subsurface samples and perform lab analysis on Mars on minerals, samples of water, and CO2, look for indications of life or complex organic compounds. (The laboratory could be pressured with a small airlock and will be teleoperated via its small robotic arms and cameras. It would use a specialized version of the virtual reality simulation of operation used by the robotic rovers. (The laboratory would use standard laboratory equipment operated in the Martian atmosphere, using the laboratory robotic arm (instruments would include rock crusher, solvents, microscope, mass spectrometer, and fluidic on-chip chemistry assays and various cultures)
- The rovers would be able to go back to interesting locations identified by study results for a deeper study
- The expedition would identify locations to support future missions for building future infrastructure.
- The equipment included would start setting up the infrastructure for Mars future returning expeditions.
- It would also provide the means for data to be collected and returned to Earth on future missions.

Most Important Requirements for Exploration and Science. What information does the scientific community need? What about Martin explorers, and the engineers who will work on this expedition need for setting the stage for future missions?

These are some of the most important things they will need to gather.

- Look for signs of life and try to establish proof
- Find (and try to retrieve) usable amounts of water and test analyze it
- Try to understand the history and geology of as large an area as possible
- Test making actionable amounts of clean water and oxygen from local water sources

- Find and prepare landing sites for future missions
- Find and prepare sites for possible future human landings and habitats
- Complete the mission objectives within a 1-2-year timeframe to be ready for the next Earth to Mars window

Hardware and Software Design objectives

The requirements for the various robotic devices and tools must as a minimum meet the following requirements.

- Robotics rovers must be extremely general purpose and have limited AI (for repetitive tasks and oversight of human VR commands) and be able to be teleoperated remotely from Earth. Functions to be controlled include roving (propulsion, batteries and power management, w backup solar panels), picking up and manipulating (dual robotic arms), communications (mesh & sat net), control of a front loader, and a sensor suite (LiDAR, cameras, ground-penetrating radar, robot gripper w/tactile feedback and acceleration measurement).
- All robotic systems and devices must be able to withstand the environment on Mars, be fault-tolerant and redundant using both NASA and Robot Battle Bot best practices. Every device must have at least one redundant duplicate.
- All devices must be able to last, be able to be repaired and, be maintained to support future missions with a 15-year target life
- Robotics must have the flexibility to meet unforeseen issues and handle unplanned failures and surprises
- The proposal must limit the complexity and the number of unique designs and devices to manage cost, schedule, and complexity.
- Together everything must meet the size and weight objectives of ten metric tons and a 5 by 5 by 5-meter cube, must survive the journey in storage, and function after deployment. Survival must address G forces, rocket vibration, and landing stresses, low-temperature extremes, two years in storage, and space and Martian radiation.
- The equipment must then survive the Martin environment of low temperature, periods without solar power, dust, winds, and wear and tear from the mission.
- A fault-tolerant process for unpacking the storage container contents must be provided

The requirements for power to sustain the expedition and to provide infrastructure for future missions must as a minimum meet the following requirements.

- Must provide 40 - 80 kWh of power each mission day to power and recharge rovers and devices.
- A large amount of power solar power is required. (~Twenty arrays 1.2 by 4.5 meters)
- Large wind turbine producing 2-10 kW depending on wind.

- Both the solar arrays and the turbine must have the ability to withstand wind, dust devils, and storms
- The power source must provide charging capabilities for robotics, laboratory, and communications
- The solar panels must have the ability to be cleaned of accumulations of dust
- All expedition equipment must be able to be relocated if required

Credibility of the Designs

What makes this proposal extremely creditable is that it does not require any new concept development for hardware. All hardware is based on proven design concepts used in new innovative new ways.

Listed below are the major points proving the credibility of this proposal:

Design concepts are based on solid engineering developed by an experienced engineer using requirements derived from scientists, NASA, ESA, Roskomos, Boston Dynamics, and Sony and using the 3d battle bot de signs provided by Solidworks for inspiration designing extremely rugged robotics

- <u>Use what has been learned in existing successful missions</u> and then use only low-risk existing technology and hardware using off-the-shelf materials.
- <u>Innovations are based on software and using proven hardware</u> to overcome major obstacles including communications lag affecting teleoperators, the complexity of communications with the Earth supporting multiple moving robotic devices, sufficient power, and communications for multiple simultaneous activities
- Using <u>generalized equipment capable of many different scientific and engineering objectives</u> and with the ability to be repurposed to address new opportunities as they occur. This allows for reduced design activities and the construction of multiple copies of hardware takes less production time to develop processes, manufacture, and debug issues
- <u>Costs and schedule is constrained using commercial off-the-shelf</u> non-exotic hardware, using a limited number of simplified designs, and the reuse of proven NASA and battle bot design concepts
- Because the overall mission needs are segregated into several independent device designs, <u>Separate projects with only limited integration</u> will allow multiple engineering, production, and testing teams to attack separate portions of the expedition designs and construction simultaneously

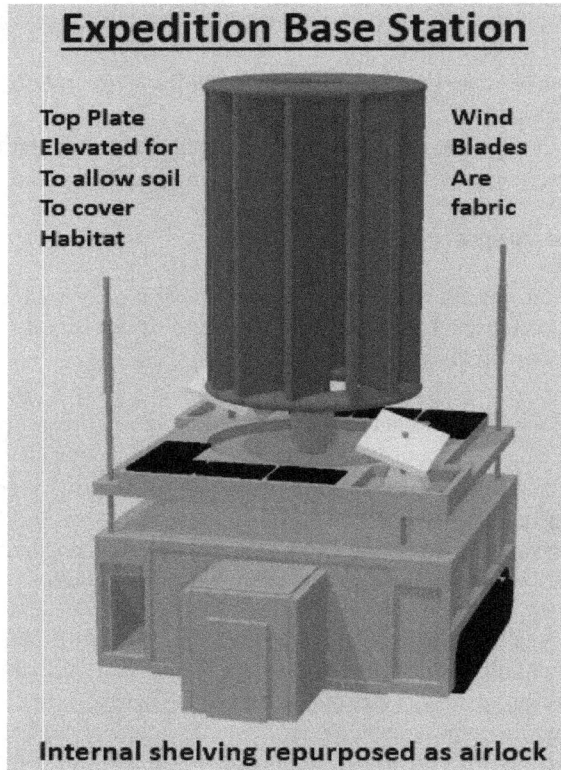

Expedition Base Station

Top Plate
Elevated for
To allow soil
To cover
Habitat

Wind
Blades
Are
fabric

Internal shelving repurposed as airlock

Figure 5 Mars Station prepared for use as habitat

The Scientific Value of this Expedition

Because of the highly generalized nature of the proposed capabilities, many varied scientific returns are planned, and importantly as discoveries are made new objectives can be added while the mission is in progress. Major scientific advancement expected in planetary geology, astrobiology, and astrochemistry including:

Detailed Geological Survey

This equipment will provide a survey covering important features, many in rough topography like the Hellas Quadrangle, where the possibility of water, life, caves and lava tubes make multiple scientific objectives possible within the range (hundreds of Kilometers) of this expedition equipment. This will support research into Geology, future site identification and exploration. Detailed mapping improvements are expected via merging existing satellite and current rover imaging with this expedition's robotic rover LiDAR, imaging, and ground penetration radar. Picking up specific rocks and samples (including drilling for subsurface samples) breaking samples open and when indicated performing laboratory mineral analysis on Mars extends these capabilities also supporting the examination of the entry to near-surface lava tubes and caves. Once-promising sites are identified core samples will be drilled at many locations.

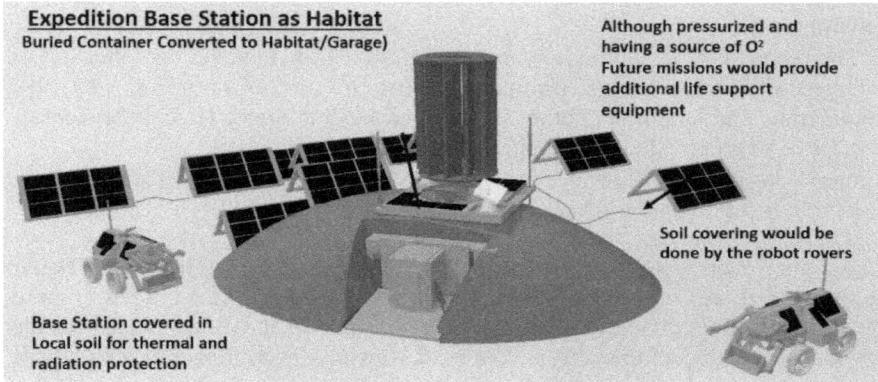

Figure 6 Mars station converted into a future habitat

Finding, identifying, and collecting potential microbial samples (including subsurface samples) and performing analysis including growth cultures, microscopic examination, and lab on chip essays (it is highly desirable that if life is found we will need to know what we are dealing with before returning it to Earth)

• Use of fluidic lab on a chip testing would be available (with hundreds of chips available optimized for use in the Martian atmosphere)

Finding subsurface water (Hydrology) and solid carbon dioxide will be done in two ways. First ground-penetrating radar used both to see where the samples are then, by drilling, sample cores will be returned for testing at the Mars station science lab using any or all the lab equipment listed below performed by an Earth based teleoperator.

• Sample crusher and prep including robot arms to move samples through tests
• Mass spectrometer
• Optical microscope
• Lab on chip devices
• Culture media

Look for signs of life and be able to establish reasonable proof. (Astrobiology). The plan is to recover and test hundreds of samples collected from promising locations via robot arm sample collecting, on multiple robot rovers, in vastly different locations (sometimes using drilling or a small pickaxe). Samples can come from core drill samples from depths up to four and one half meters.

Preparing for Future Human Exploration

The preparation for future crewed exploration has three elements, finding future sites of interest, leaving useful infrastructure, and better understanding the Mars environment.:

- The preparation for future human exploration will include the identification of future sites for landings. Habitat construction, sources of water, CO^2, key minerals, and targets for future exploration landing areas, landing beacons, a store of water & O2, and a significant electrical power source.
- Improving knowledge of any potential Mars biome to better protect future human explorers, Earth, and the Martian environment.

<u>Exploration survey of potential sites for landing and future habitats</u>. Habitat locations that are sheltered from wind and radiation would be explored in more detail. These are cliffs, lava tubes, caves, and large rock structures. Also possible are locations adjacent to sources of water. Likewise, flat landing zones with hard exhaust resistant terrain near potential habitat locations would be located. Also, important will be leaving (after this exploration mission) power, Habitat, communications, and robotic excavation and construction capabilities for future scientific missions to perform tasks not yet envisioned.

<u>Robotic tools to support scientific study</u>

- Robotic rovers -_Rovers equipped with ground-penetrating radar and acoustic sensors and would be able to precipitate in subsurface mapping by performing laboratory examinations for impurities and testing various methods of filtration and purification
- Using a wheeled vehicle with the ability of two robot arms which can also assist in climbing or descending through extremely rough terrain and by deploying mesh network communications repeaters allowing for operation and penetration of near surface caves, lava tubes, etc.
- The combination of a very robust rover unitizing LiDAR, optical imaging, subsurface radar, and acoustic mapping, allows the detailed exploration of near surface caves, lava tubes, and overhanging cliff areas not mapped by satellites. Materials analysis will be able to collect a vast amount of new data that would enhance the understanding of Mars.
- The very robust robotic rovers are designed for traveling more dangerous and interesting terrain than current rovers and have more in common with a Earth bobcat than current rovers.
- Using the piton's explosive charge would also allow acoustic sensors in multiple robots to map the subsurface structure.

Boring Drill

- 4.5m Remote controlled speed & down pressure
- Deployable support legs
- Internal rechargeable batteries
- Dimond rock cutting bit w/ sample retrieval in bit

Smart Piton

Mini Winch
- Retract cord with hook on demand
- Also used on solar arrays

Explosive Charge
- Robotic rover holds with gripper
- Secure piton on command, bang
- Also, can provide
- impulse for subsurface mapping

Sample retrieval bit

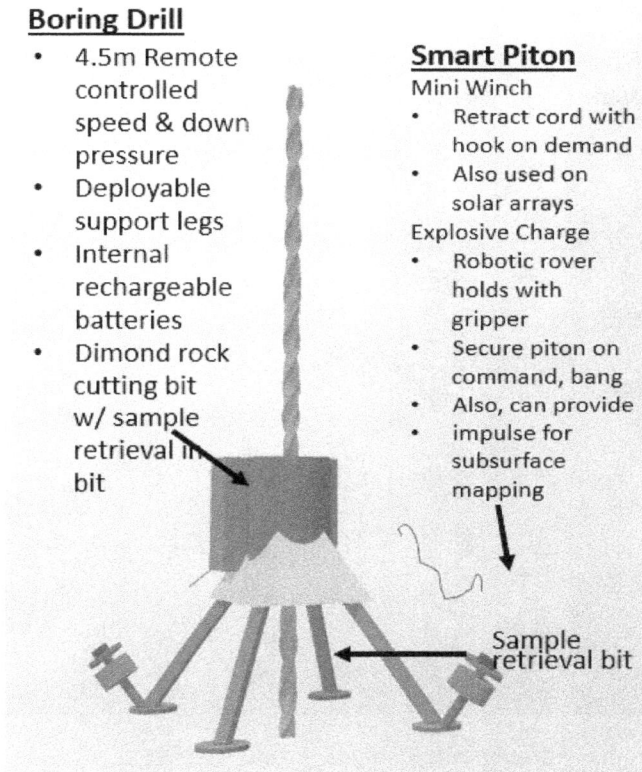

Figure 7 - Sample Drill & Smart Piton

Sample drills would be provided for drilling to depths of up to 4.5 meters. These will utilize a design optimized for use on Mars. After being attached to the ground with explosive pitons deployed by the robotic rovers an electric motor will spin a cutting head that will slowly bore into the surface cutting rock and hard soils as it proceeds. The drilling devices, deployed by any one of several duplicate robot rovers as suitable sites were identified. The drill head would be hollow and designed to capture a sample from the maximum depth drilled. After visual examination promising core samples would be returned to the Mars station science lab for deeper laboratory analysis including looking for subsurface, water, carbon dioxide, organisms and detailed material composition

Robotic Rovers: The rugged design of the robotics presented is ideally suited for rapid exploration of difficult and rugged terrane over large areas including lava tubes and potential areas of surface water and frozen CO^2. Each rover will be able to move quickly, climb steep embankments, self-right if overturned, tow other rovers, transport devices, plow or dig soil, and will have two robotic arms for manipulation of tools, tie-down pitons, connecting power cables, or for setting up solar panel arrays and subsurface drilling rigs. (or any general construction)

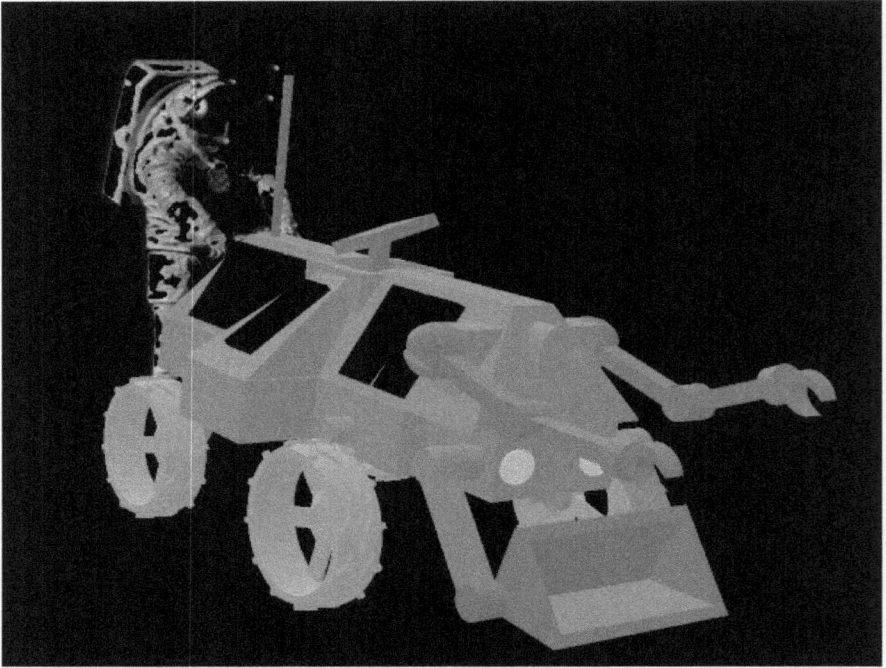

Figure 8 - Manual operation of rover on a future mission

The robotic rovers presented will have the ability to perform light excavation of sand and local soils as well as driving pitons into the sand, soil, and rock to secure equipment from winds and the ability to do basic construction work via a teleoperator. landing sites and habitation sites can be cleared and flattened. Sand and debris can be cleared for future missions for the lifetime of the robotics. The robotic rovers used in this mission would still be available to support future missions after this expedition is completed. They also have a footstep and manual controls on the back to allow them to transport and be operated by astronauts in space suits in a standing position during future crewed missions. All equipment is general-purpose, robust, and has significant duplication. The infrastructure left in place by this mission will be highly useful for future missions supporting additional exploration, lab analysis, supporting the offloading of a future mission's equipment and construction of future infrastructure.

Mars Station & Laboratory. Robotic Shipping Containers are transformed into Mars stations which then transform into a habitat. Each of the two 5 x 5 x 2.5-meter shipping containers would become Mars stations supporting this expedition and they and their associated solar power arrays would be relocated near future habitat sites to provide a significant source of habitat, power, water, O^2 and laboratory services to support future mission requirements The containers would be mostly empty space after the robotics and tools were deployed and would be designed for reuse and be able to hold Earth standard atmospheric pressure. The container would retain large battery packs, power management, wind turbine, the laboratory, and mars Earth communications and antennas. Solar arrays would be deployed and connected to the Mars station via cable and would be secured to the ground using pitons. Radio beacons installed on the transport containers would improve the safety of future landings in all weather. All of this would be left in place near a future habitat/ landing site and would provide significant infrastructure.

The design of the containers will allow the communications, wind turbine, and a small set of solar panels to be separated from the habitat allowing the main portion of the container (containing the laboratory, batteries, water and LOX production to be buried as protection from radiation, extreme temperatures and storms. The expedition would provide two habitat shelters each with fifteen square meters (30 cubic meters) of future habitat space. The internal shelving and bracing needed for the voyage removed and reconfigured into an airlock and internal furnishings

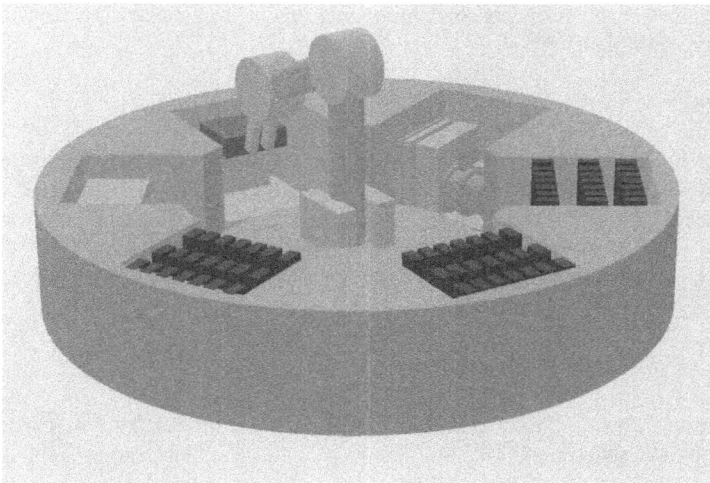

Figure 9 – Carousel of laboratory equipment teleoperated by robotic arm

The containers would contain a telerobotic laboratory that would utilize a suite of typical laboratory instruments, a rock crusher, solvents, culture mediums, and a collection of lab-on-chip devices. A robot arm will access various science stations in a carousel, A scientist / operator on Earth will setup and run analyses and experiments using the same VR/Digital twinning process as used by the rovers and then wait for the results to be transmitted to Earth allowing experiments to be designed based on what is being discovered. This also leaves in place significant laboratory capabilities for future missions. Additionally, experiments will be included. Such as pilot equipment to produce quantities of clean water and liquid oxygen made from Mars local water sourced as ice supplied by the rovers and then melted filtered and electrolyzed in the Mars station.

Credit goes to an experimental oxygen production system developed by the University of Wollongong on their Capillary-Fed Electrolysis (CFE) innovation providing an extremely power-efficient electrolyzer design with 98% energy efficiency (per the scientific journal Nature).

This is an experiment and would be targeted to produce two liters per day of oxygen (on days where enough solar power is available) and store water and LOX in temperature controlled insulated tanks to support future missions. The design target for the expedition onboard O^2 generation experiment (using water mined on the surface and melted and filtered to be electrolyzed) is (871g O^2 per kg of water using 47.5kWh/kg. while using about 2kg of water per day)

This resource could provide oxygen for fuel cells or provide emergency oxygen reserves for future crewed missions (but not enough for rocket fuel) and would set the stage for larger more industrial scaled equipment needed for future missions. A source of oxygen alone does not provide life support - additional provisions required in future missions for managing CO^2 and balancing the artificial atmosphere composition.

The Expedition Cost

Listed below is the hardware and software needed to support this expedition. Each has a development cost, a production cost to physically produce the hardware and an ongoing operational cost. I have estimated the development and production costs, assuming that the operational costs will be carried in the overall mission costs. These are the costs for a university team not a NASA contractor, which would be much more)

Earth based software required will include:

•   Mission Command (Rover Teaming VR, Software for tasking mission elements) - software

•   Mars station operations software
    •   VR for rover mobility and sensor management – software
    •   VR for Mars station mobility and sensor management - software
    •   VR for Laboratory operations
    •   Software to manage communications and power management -software
    •   Software to remotely operate the Water / O^2 pilot – software

Mars based Robot Rovers will require both the physical hardware and specialized software. (8 included in mission, four per station)

- Rover design and construction includes propulsion, navigation, robotic arms, and sensors – hardware
  - Basic Rover design
  - Sensor Suite
  - Propulsion and guidance and Mars station VR
  - Communications systems (Earth sat comms and mesh network)
  - Robot arms
  - Solar backup, batteries, charging, and power management
  - Onboard systems management and AI software

Mars stations (2 included in mission). Both will require both the physical hardware and specialized software.

Basic Container, Mars station, and Habitat shell includes propulsion, navigation, sensors, and internal transformable partitions
- Communications systems (Earth sat comms and mesh network)
- Laboratory and Laboratory onboard management and AI
- Wind Turbine, on station solar backup, batteries, Charging and remote solar power management
- Water and O^2 production and storage pilot hardware
- Propulsion and guidance hardware and sensors for rovers

Additional tools custom designed for expedition
- Solar Panel Arrays (20ea., ten per station)
- Boring/Drilling Rigs (6ea. 3 per station)
- Small Tools -Pitons (40ea. 20 per station)
- Small Tools -Pick Axe (4ea. 2 per station)
- Small Tools -Panel Broom (4ea. 2 per station

Exploration Mission Cost Estimate

The assumption is based on an accelerated build using costs based on university graduate students designing and building all aspects of the project

Please note that these costs are likely unrealistic in the real world,

Table 1 -Estimated Cost of All Expedition Elements

			Total cost estimate		$	23,535,441.00
Earth command software and stations required			Person-hours	Materials	Cost	
•	Mission Command		8300		$	1,228,400.00
•	Mars station Operations		8394		$	1,242,312.00
Robot Rover Operations (Rover specific VR)						
•	Science Advisory Team		1129		$	167,092.00
Mars stations/Habitats (2ea.)			Person-hours	Materials	Cost	
•	Basic Container, station, & Habitat		15939	$ 649,392.00	$	3,008,364.00
•	Sensor Suite		5043	$ 14,939.00	$	761,303.00
•	Propulsion and guidance		7493	$ 43,922.00	$	1,152,886.00
•	Communications systems		5939	$ 49,220.00	$	928,192.00
•	Laboratory and Laboratory VR		11920	$ 124,382.00	$	1,888,542.00
•	Power and Power Management		4983	$ 21,994.00	$	759,478.00
•	O2 production and storage		7372	$ 12,034.00	$	1,103,090.00
•	Internal transformable storage partitions		1293	$ 9,932.00	$	201,296.00
Robotic Rovers (8 total, 4 per container)			Person-hours	Materials	Cost	
•	Basic Rover design		12082	$ 294,328.00	$	2,082,464.00
•	Sensor Suite (shared with Mars Station)		4939	$ 17,039.00	$	748,011.00
•	Propulsion and guidance and Mars station VR		5943	$ 384,832.00	$	1,264,396.00
•	Communications systems		12920	$ 84,932.00	$	1,997,092.00
•	Robot arms		8504	$ 29,329.00	$	1,287,921.00
•	Power Management		8695	$ 32,999.00	$	1,319,859.00
			Person-hours	Materials	Cost	
Solar Panel Arrays (20ea., 10 per container)			5939	$ 54,900.00	$	933,872.00
Boring/Drilling Rigs (6ea. 3 per container)			4922	$ 13,920.00	$	742,376.00
Small Tools -Pitons (40ea. 20 per container)			3822	$ 12,999.00	$	578,655.00
Small Tools -Pick Axe (2ea. 1 per container)			493	$ 5,939.00	$	78,903.00

How Soon Could the Mission be Ready and How Credible is a launch by 2033

To address both the timing and the creditability the draft program schedule shows that both the development timing and the production schedule are achievable within the 5 years allowed to develop and build the expedition robotics, software and tools needed. The chart shows that each of the individual tasks are achievable within the allotted time. (See following schedule chart)

Expedition Timeline

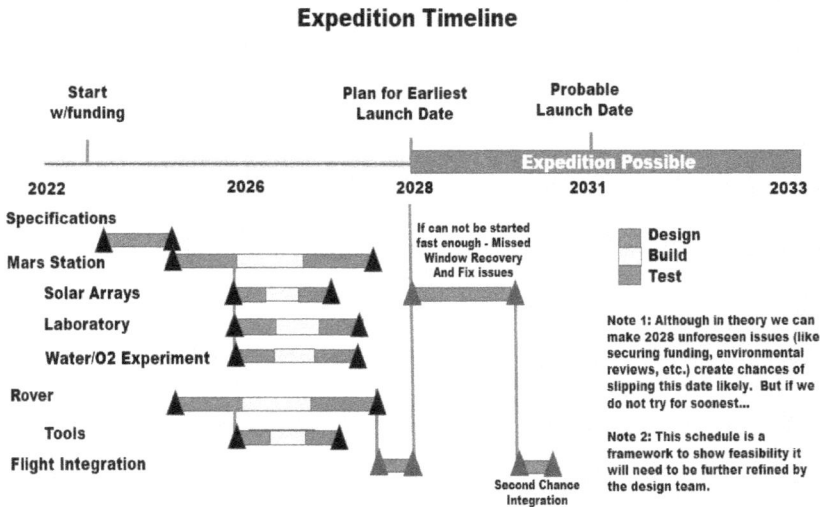

Figure 9 - Proposed schedule plan for expedition

It is required to use multiple design, production, and testing teams to separately attack the major design elements. (The separate projects include the two-shipping container/Mars stations, the eight robotic rovers, the solar panel arrays, the boring rigs, and the small tools) All of the design requirements are designed to accommodate rapid low-risk development,

The innovations proposed (especially the use of a virtual reality Martian environment to work in) are software and can be developed and used within the development timeframe even though they would enable vastly more science and significantly more future mission preparation than could be envisioned without compensating for the significant time delays.

The hardware is based on current technology, and it is possible that it can be developed in the timeframe needed.

All designs are based on existing robotic concepts and designed using conventional materials and built with standard manufacturing practice

How much earlier could we launch?

The answer is withing 4 years, this would not likely achieve an early launch window (because of the 2-year launch cycle and would increase risk. Even the 5-year schedule proposed will require an extreme effort.

The answer accelerating the schedule depends on many uncontrolled variables.
- Is there the political will to use SpaceX style development with risk of failure?
- Is the funding and an organization to drive the program ready to start day one?
- Is the economy in a state where people with the right skills are available immediately.

Technical Design Details

Many important details are shown in the following section.   These are not a complete set of requirements, only some of the more important requirements that will fit in a 20-page proposal.

<u>Volume and Weight</u>:

Listed below is a table compiling the reasonable estimated weight targets for each sub-item demonstrating that the sizes and weights are feasible.

Table 2 -Estimated size and weight,
showing that design can reasonable meet requirements

Volume and Weight Allocations	Item Volume in M³	Volume Of station	Volume of 2 stations	Item Weight in Kg	Sub	Weight in Kg for both
Target for station (2ea.)		**62.5**	**125**		**2500**	**5000**
>>Sensor Suite	0.5	0.5	1	30	1	2
>> base station VR	2.6	2.6	5.2	237.5	237.5	475
>>Communicatio ns	1.4	1.4	2.8	40	40	80
>> Laboratory VR	7.1	7.1	14.2	162.5	162.5	325
>> Power Management	13.2	13.2	26.4	182	182	364
>>O² production and storage	2	2	4	90	100	200
>>Transformable partitions	1	1	2	75	75	150
Robotic Rovers (8ea.)	1.9		7.6	332.5		2660
>>Sensor Suite	0.5	3	6	30	120	240
>> Rover operator VR	1.2	7.2	14.4	85	340	680
>>Communicatio ns systems	0.7	4.2	8.4	35	140	280
-Robot arms	0.9	5.4	10.8	56	224	448
>> Power Management	1.4	8.4	16.8	110	440	880
Solar Panel Arrays (20ea)	0.15	14.5	29	13	130	260
Boring/Drilling Rigs (4ea.)	1.4	8.4	16.8	90	180	360
Small Tools - Pitons (40ea.)	0.3	6	12	5	100	200
Small Tools - Pick Axe (2ea. 1 per container)	0.1	0.2	0.4	15	15	30
Small Tools - Panel Broom (4ea. 2 per container)	0.16	0.32	0.64	8	16	32
Total		85.42	178.44		2503	5006

Important Requirements

This section provides additional information needed to understand and specify the development needed to make this proposed expedition a reality:

- All technology proposed uses standard commercial electronics, 3d printed plastics, and conventional construction with aluminum and stainless steel to reduce the design and fabrication costs
- Using electrostatic force to cause dust particles to detach and virtually leap off the solar panel's surface, without the need for water or brushes. To do this a row of wires (above the panel) is charged, in sequence. above the solar panel's surface, imparting an electrical charge to the dust particles, which then repels the dust.

Electrostatic dust removal is inspired by MIT graduate student Sreedath Panat and professor of mechanical engineering Kripa Varanasi. Despite concerted efforts worldwide {and on Mars} to develop ever more efficient solar panels, Varanasi says, "a mundane problem like dust can put a serious dent in the whole thing." They have developed and tested a simple process to remove dust from solar panels by using electrostatic charge.

Providing for a swarm of individually controlled robots requires significant local badwith, the use of Mars based mesh networks eliminates much of the complexity of the individual units. Only the 2 Mars stations need redundant high bandwidth communications to Earth and Martian satellites. The rovers will carry a minimal low bandwidth satcom to Earth capability for backup and out of range use.

- The use of some human command and control reduces the requirement to develop costly AI. This also allows multiple human operators to manage many different explorations and projects simultaneously. This is only possible because of the ability to use VR and digital twinning to vastly reduce the bandwidth needed and to provide operations that do not seem impacted by communications delays.
- One of the requirements of making exploration much faster Is that individual robots will not be able to accumulate sufficient power with self-contained solar panels. This requires that they return to the Mars station to recharge batteries. They would have self-contained solar to allow communications and a very slow return if for some reason batteries were depleted before returning for a recharge
- The laboratory will need to include ruggedized mass spectrometer microscope, tunneling microscope, fluidic lab on-chip dispenser and storage, and a small robotic arm to remotely operate all equipment. It will also have a pressurized compartment with a small, airlock
- The design of the robot rovers is based on a four electric powered wheel all-terrain chassis with two robot manipulator arms, batteries a small backup solar panel, and electronics. Attachments for the arms include a shovel, a small pickaxe, a solar panel cleaning squeegee, and lab sample collection containers

for transferring samples to the Mars station. The arms can also grasp the core drill and the explosive pitons. The arms can be used to right the robot if overturned.

- The rover chassis also has a frontend loader-style bucket with a hook that can be used as a plow to lift and move soil or to hook other equipment for towing.
- Rover Sensors include LiDAR, Cameras, acoustic sensor, gyro, accelerometer, tilt sensors, ground-penetrating radar, and tuned microwave sensor for detecting water and separately dry CO_2.
- The design of the core sample boring drill provided for drilling up to depths of up to 4.5 meters. After attachment to the ground by pitons an electric motor will spin a cutting head that will slowly bore into the surface cutting rock and hard soils as it proceeds
- The design of remote solar and battery-powered mesh repeaters would be small expendable devices with a small integrated solar panel that would extend the range of the shared mesh network they would be placed at the entrance to caves, at the top of cliffs, or anywhere the mesh network signal became weak to assure that the robotics would maintain robust communications.
- The design of the Solar panel arrays would be like the panels used on Earth (except lighter, thinner, producing about one-half the output per area of Earth-based panels, and would be designed to withstand the sandblasting of wind) Panels would be pinned down to the ground using the smart pitons with dust cleaning using either electrostatic force or by dry sweeping using a special broom via the rover robotic arm.
- The design of explosive pitons (used for a variety of surface attachments) would be held by a robot arm and either driven into the ground by pushing or driven by sending arming and firing commands to detonate a small explosive planting charge. They would imbed themselves into the sand, soil, and low-density rock and be expendable. Once embedded they would retract a cable on command

Ideas from outside the limits of this challenge

- A Nuclear power source for the base stations
- A constellation of Martian communications satellites
- A few relay satellites between Earth and Mars to reduce the effects of solar occlusions
- Perseverance style helicopter to extend the range of surface modeling and plant repeaters
- Multiple duplication of the telerobotic mission to increase surface coverage with a larger number of rovers
- The concept of heavy lift robotics expedition lends itself to the creation of a standard exploratory package that can be produced in volume and used in multiple launches, exploring multiple locations on Mars. (or with variations the Moon and other deep space destinations.)
- We need to use the best ideas from all parties without ego and that includes help from NASA and ESA who have a wealth of experience.

CONCLUSION

This design proposal provides a framework that can meet the requirements of this challenge. It will allow rapid development of a capability that is extremely flexible and can support not only the objectives stated in this paper but many other objectives yet to be understood. The most important innovation is the use of digital twining and virtual reality to allow more work to be done more quickly. I hope this paper challenges you the readers to use the ideas presented, not just in this paper, but the best ideas from all the submittals for the advancement of planetary science and the betterment of humankind.

I want to end saying the most important result of this design challenge is that the best ideas from all contestants, become adopted.

I am honored to present my ideas to contribute in a small way towards adopting Mars as our second home.

2: PROJECT LATE (LAva Tubes Explorers)

Innspace Team
Piotr Torchała, Justyna Pelc, Beata Suścicka, Ewa Borowska,
Magdalena Łabowska, Nikola Bukowiecka, Aleksandra Klassa, Bartosz Rybacki,
Hubert Gross, Nikodem Drąg, Michał S łomiany, Kamil Ziółkowski,
Kajetan Szostek, Milena Michalska
innspace@innspace.pl

The main question that humanity is looking for an answer is: Are we alone in the universe? Finding life on the surface of Mars is highly unlikely. Conditions on Mars are not conducive to survival. Large fluctuations in temperature, radiation, and dust make finding life difficult. We need to find a place characterized by stable conditions - it is best to go underground. Drilling into the Martian surface is not a simple task, as we already saw in the Insight mission. The Innspace Team decided to look for natural spaces whose construction protects them against external conditions. These spaces are lava tubes.

LAVA TUBES

The lava tubes on Mars are the remnants of flowing lava that form underground tunnels with a circular or arched shape. Similar lava tunnels can be found on Earth, for example, in Hawaii. Due to the lower gravity on Mars, which is only 38% of the Earth's, the lava flowing on Mars could create tunnels with a larger cross-section than on Earth. Martian lava tunnels have yet to be explored, and little is known about their geology and formation. Their presence was discovered thanks to the craters created by a meteorite impact that filled the tunnel's ceiling.

Lava tubes can be an ideal candidate for the search for life. Down the tunnel, the temperature should be constant, and its fluctuations should be small. A thick layer of regolith protects the tunnel environment from radiation. That's why life had a better chance of survival, and the chances of finding it by sent robots are higher. They are an exciting exploration target, not only in terms of science but also in terms of future Mars exploration. They are also considered a potential location for building a base on Mars. In addition, there may be frozen water where we can look for signs of life; in the future, it can supply the base with this element necessary for survival. We can provide answers to crucial questions bothering humanity while preparing for a crewed mission to the Red Planet. The study of lava tubes, however, is a critical element in the further human exploration of Mars.

Why are lava tubes scientifically fascinating?

The lava tube environment represents the primary environment, which allows us to conduct research related to the search for life or its traces. Due to temperature and pressure fluctuations, specific geological processes and chemical modifications of minerals in rocks form various rock formations. Thanks to the newly formed deposits and minerals, abiogenesis and the emergence of inhabited micro-worlds can occur.

One example of such a phenomena is the transformation of peridotite into serpentine. Lava tubes offer a potentially wide variety of microbial environments. Due to the geological structure of lava tubes, there is a possibility of the presence of lithotrophic and endolytic organisms. The research we are proposing also broadens our current knowledge of lithotrophic microorganisms, mainly how they contribute to releasing biogenic elements inside micro-niches that stimulate a prosperous life. These microorganisms crush rocks/minerals containing a phosphate anion such as apatite, vivianite and other bioavailable inorganic and organic compounds. This is the reason why these are the elements we are looking for. All the conducted experiments help us discover and model how the first organisms influenced micro-environments in the early stages of the Earth's development and how micro-environments influenced the origins of life. This helps us understand how we can survive on Mars and other exoplanets.

How will it meet our plans of sending a crewed mission to Mars?

In addition to looking for life, our mission is to prepare for the first crewed missions. The concepts of the first Mars bases consider the creation of such a colony in the lava tunnels. Preparing tunnel maps will allow us to prepare for creating such a base in these tunnels, and learning about the composition and construction of the tunnels will allow us to check whether these tunnels are sufficiently durable for this purpose. Let's not forget about water - finding it in lava tubes would be of great importance for the success of the future mission, and there is a high probability that we will find deposits of it inside lava tubes.

Conclusion: our chosen location is the Coprates Canyon area

We decide to choose the location where we suspect to find a large number of lava tubes (existing of some of them is already confirmed), in which there are also other areas of interest to us. The mission's target is Coprates Canyon, located next to Coprates Chasma. It is part of the Valles Marineris canyon system.

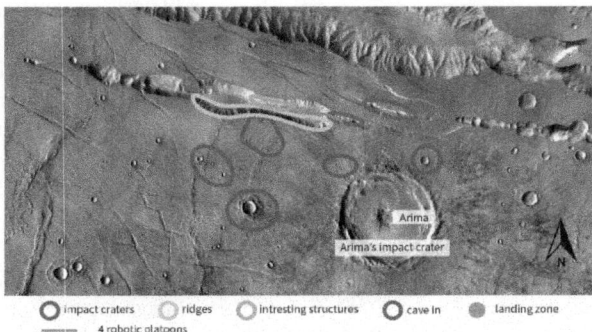

Figure 1 Coprates Canyon

The canyon is easy to exploit, and slight differences in altitude are confirmed by the data from the JMARS program presented in the pictures below. You can see that its depth is about two kilometres. Its slope at the steepest parts is about 30%. The robots currently in use on Mars are vehicles capable of 45%, and for safety, it is limited to 30% for the duration of the mission. This canyon combines many variants of sinkholes.

The satellite photos show the chains and ridges. They are mainly formed when the lava tubes collapse so we can assume the presence of the lava tube entrances. In the canyon and Lava Tubes, there may be found such materials as pyroxenes, iron phyllosilicates (along the walls, in the walls - in many places), phyllosilicates, calcium pyroxene, peridotite, serentynite - significant in that they contain magnesium, iron, aluminium, silica, apatite, as well as sedimentary rocks/sedimentary minerals (clays, solids). In addition to the Lava Tubes in the canyon and the surrounding area, rampart craters indicate surface volatiles/ground ice or, in other words, water. Valles Marineris is believed to contain relic ice in regolith - which is extremely important for constructing a future Mars base. There are mud vulcanos in the area, which are very attractive in terms of mineral exploration, and there are organic salts that can be helpful in oxygen extraction. We also have many young impact craters that are important for crater counting. In addition, on the edges of the canyon, we can also find craters, ridges and sinkholes.

Figure 2 Examining the slopes - selected paths

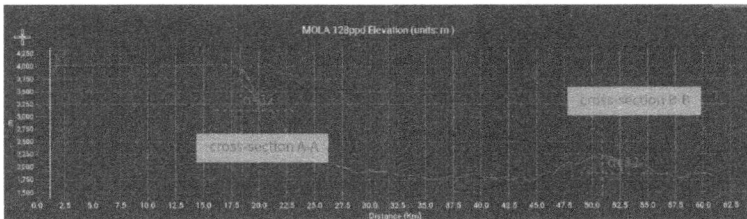
Figure 3 The plot of the slopes of the paths marked in the picture above

That's we decided to split our mission into two parts. 2 platoons examine the lava tubes inside the canyon, and two platoons stay at the top of the canyon to explore mentioned areas (craters, mud vulcanos) in terms of looking for life and preparing for a crewed mission. This way, we can examine the more extensive area and minimize the risk of determining Coprates Canyon as unhabitable for the future crew.

The landing takes place next to the crater. We do not have data on the landing ellipse of the lander, so we cannot determine the possibility of landing in a crater to speed up the mission. If the lander could land with the same accuracy as the Perseverance rover lander (7.7 km x 6.6 km), we would be able to land at the bottom of the crater because it is at least 8 km wide.

GET TO KNOW PROJECT LATE

The challenge is to propose a mission with a substantial scientific output that answers the critical question about life on Mars, designate a site for a crewed mission and make the first preparations for it. The considerable challenge is the limited payload, the significant risk of failure, and the still little information about Mars and what awaits us. Robots are a crucial component of crewed missions. Before we send a human on such a long and dangerous journey as the flight to Mars, we must carefully examine this planet and prepare the appropriate conditions for our crew. To set foot on Mars anytime soon, we can't ship one robot every few years. We also need a perfect reason to mobilize governments worldwide to allocate significant resources to such a mission, as no country can do it on its own.

That reason would be to find life. Such information would put the whole world on alert. Therefore, we must move to places with the best living conditions and search as large an area as possible to increase the likelihood of breakthrough discoveries. We have to go underground. Lava tubes are a promising area, but completely undiscovered. We do not know what to find in them and how to select the best caves for research. That's why we proposed the LATE (LAva Tubes Explorers) mission with 4 riding, 24 walking and 116 flying robots.

To carry out a complex operation in a harsh environment on Mars, it is necessary to rely on the diversity of robots in selected platoons and sections, as well as redundancy. To explore a large area of terrain quickly, it is necessary to have many robots. There are 4 robotic platoons in the lander, whose leading units (commanders) are large mobile transporters. This platoon consists of five types of robots:

- transporter rovers - riding robots based on Curiosity and Perseverance, responsible for delivering the walking robots to the Lava Tubes and supplying them with power via MMRTG / Kilopower (to reduce the number of necessary reactors = lower total weight of the robots),
- walking robots - adjusted Boston Dynamics Spots, responsible for scouting the lava tunnels and creating 3D maps of lava tubes,
- two types of flying robots - rotorcrafts and tilt-rotor aircraft, responsible for reconnaissance and sample transport,
- Minibots powered by small jet engines explore narrow passages.

TRANSPORTER ROVER

The transporter rover serves as the platoon leader. Its job is to transport and charge Spots from the walking section, transport Minibots, transport collected samples, drill boreholes, minor service malfunctions in the robots and communicate with the lander. The transporter rover consists of components with a TRL of no less than 6. Components were taken from the Curiosity or Perseverance rovers.

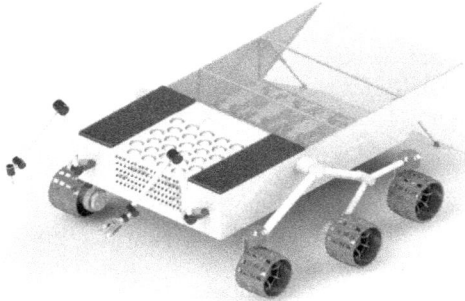

Figure 4 Transporter rover

The rover's rocker-bogie suspension was selected as reliable, tested on the Mars subsystem and scaled to meet the needs of the transporter. The rover body consists of lightweight, high-strength composites attached to an aluminium truss. The rover is powered by two Kilopower units of 1 kW each (one to charge the Spots, the other to operate the transporter). Above the Kilopowers are radiators used to control the temperature inside the transporter. On the front of the rover is a closable place for placing samples collected by the Spots, and on the top is a place for 27 Minibots. The Spots placed inside are inductively charged using a WiBotic's panel.

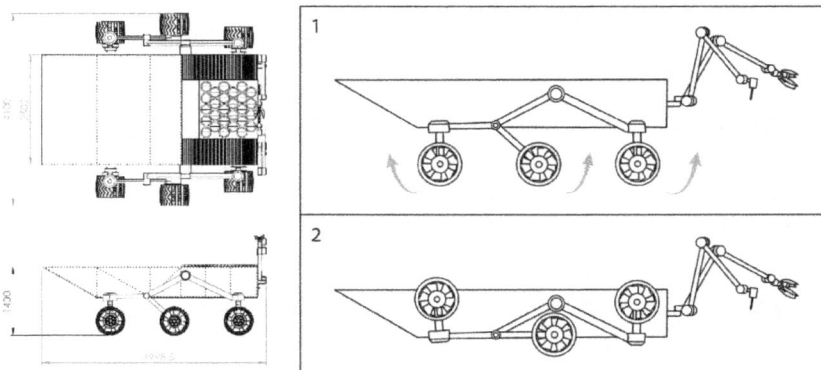

Figure 5 Dimension of the rover and stowed configuration

The power supply of transporter rovers

The transporter rover is powered by a 2 Kilopower 1 kW generator that uses nuclear fission. To minimize the mass of the generator, it was decided to use HEU (Highly Enriched Uranium) as a fuel, which can be 700 to 800 kg lighter than LEU (Low Enriched Uranium). In the case of the transporter rover, each of the 1kW-HEU Kilopower used has a mass of 339 kg.

Scientific instruments in transporter rovers

Transporter rovers are equipped with 2 key instruments: an XRF (X-ray fluorescence) spectrometer for testing the chemical composition of rocks and NIR (Near-Infrared) Imaging Cameras for assessment of the mineral composition of rocks and evaluation of specific minerals.

XRF measures basic chemistry at sub-millimetre scales by focusing an X-ray beam on a tiny spot on the target rock or soil and analyzing the induced X-ray fluorescence. The instrument consists of the main electronics unit in the rover's body, and a sensor head mounted on the robotic arm. The sensor head includes an x-ray source, X-ray optics, X-ray detectors, a high-voltage power supply, a micro-context camera, and a light-emitting diode. As a result, we can determine the presence of such elements as sulfate and sulfide search - indicating the presence of water in the past. It can also detect Na, Mg, Al, Si, P, S, Cl, K, Ca, Ti, V, Cr, Mn, Fe, Co, Ni, Cu, Zn, Br, Rb, Sr, Y, Ga, Ge, As, and Zr.

NIR Imaging Cameras is the pencil-beam infrared spectrometer that measures reflected solar radiation in the near-infrared range. The instrument works in the spectral range of 1.15–3.3 μm with a spectral resolution of 25 cm-1. The optical head is mounted on the mast, and its electronics box is inside the rover's body. The main goal is: mineralogy characterization and remote identification of water-related minerals, contributing to the selection of suitable samples for further analysis and geological research. What's more, the NIR Imaging Camera enables us characterization of the composition of surface materials, discriminate between various classes of silicates, oxides, hydrated minerals and carbonates, identification and mapping of the distribution of aqueous alteration products on Mars as well as real-time assessment of surface composition in selected areas, in support of identifying and selection of the most promising drilling sites. Thanks to this instrument, we can conduct studies of variations in the atmospheric dust properties and the atmospheric gaseous composition.

The rover also has basic sensors for determining environmental parameters and sensors for 3D mapping and movement in the field. For navigation and mapping, it uses a set of stereovision cameras with illuminators, a set of navigation cameras with illuminators, thermal imaging cameras and lidar. They create a spatial map around the rover. The sensors for studying the space around the rover are based on devices from previous Martian missions.

SPOTS

To explore a large area of Mars in a short time, the platoon of the robot dogs model was adopted. During the mission, each of the transporter rovers that provide shelter and power to the robo-dogs and Minibots explores a different place of scientific interest. Whether mapping terrain, collecting samples or taking pictures, the robots are spread out around the transporter rover.

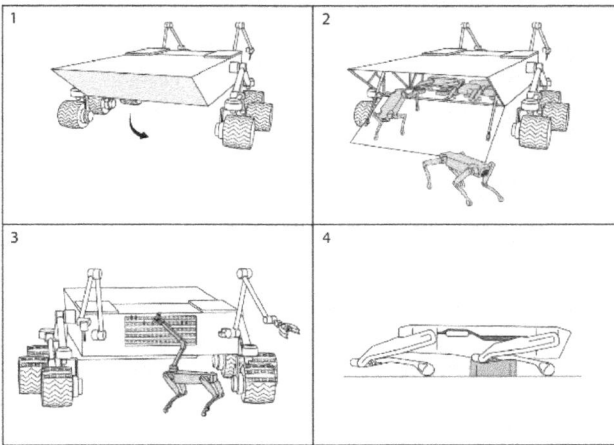

Figure 6 Transporter rover with Spots and Minibots inside

Walking robots explore caves and tunnels and aggregate data about them to create 3D maps of the studied areas. These robots have a set of sensors in the form of stereoscopic cameras with an illuminator, thermal imaging cameras and lidars. Cameras and lidars help determine the route of robots, and lidars, thanks to greater accuracy and range than cameras, create a map of inaccessible areas. The prepared 3D models of the tunnel interiors are used by robots (in real-time) to determine the route trajectory and Mission Control Center to identify potentially interesting places. They also play an essential role in future crewed exploration and building a Mars base. When a geologically interesting site is found, the rover transporter stops. It then opens the tailgate and allows the robo-dogs to explore a large area quickly. Thanks to their speed of up to 1,6 m/s and an operation time of 90 min (including 3 min to enter the loading dock of the transporter), the 6 robo-dogs can cover a distance of about 50 km during a single course.

Robo-dogs charging

Spots are charged wirelessly using the WiBotic onboard charger. Wireless charging has several advantages: no charging ports that can get clogged with dust, no protruding parts in the rover transporter, and no need for a perfect match between the transmitting and receiving coil (no need for contact). This solution has been tested in a simulated lunar environment.

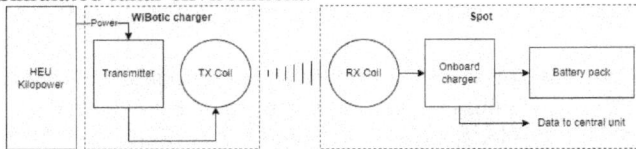

Figure 7 Charging of Spots

Communication and navigation

The communication system must meet the following requirements: the ability to transmit a signal across the surface of Mars into a cave, the ability to send a signal over a 5 km lava tube distance (terrain with a possible significant slope) and support different types of transmission modes, such as data transmission and real-time data transmission. The proposed design is a mixture of VLF Through-the-Earth (TTE) and Line-of-sight (LOS) communication. The VLF data rate is about 300 bits/s, so data compression is crucial. The system provides for the use of VLF frequencies up to 10 kHz. Natural background noise increases with decreasing frequency but is negligible in a given environment. The most noticeable noise is that of the rover's electric motors, which can be easily filtered out using FFT analysis.

Through-The-Earth (TTE) signalling is a type of radio signalling used in mines and caves that uses low-frequency signals (300-10000 Hz) that can travel through tens of metres of rock layers. The antenna cable can only be on the surface and provide coverage in the cave. The antenna is placed in a "loop" formation around the cave's perimeter for better coverage. Transmissions propagate through the rock layers, which are used as a medium to carry very low-frequency signals. Because the signal travels through the rocks, the antenna does not need to run in all parts of the cave to achieve broad coverage. Considering the lava tube's unknown shape and profile, resulting in a high risk of losing the robot that gets into explore, very low-frequency radio communication seems the most reliable and mass efficient.

To navigate, robots scan the entire 360° area around themselves and measure cave height to prevent the rover from touching the cave ceiling. During operations, if the rover loses signal from the base, it can autonomously calculate and execute a way back to the base or nearest transporter rover. The navigation algorithm would use a dead reckoning process. Spot's kinematic model is implemented to determine which route is accessible by the robots. To achieve that, Spot's equipped with thermographic cameras for imaging and mapping the inside of the lava tube using infrared radiation and initial assessment of mineral composition and lidars for making high-resolution maps.

AERIAL VEHICLES

We proposed two different aerial vehicles: tilt-rotor aircrafts and rotorcrafts. The primary purpose of flying vehicles is photogrammetry and terrain mapping. The tilt-rotor aircrafts are used for reconnaissance, their role is to explore as much terrain as possible as quickly as possible, making it possible to create accurate 3D surface maps that help in further mission planning. Rotorcrafts are responsible for geological research and a detailed exploration of distant (for the rover) Points of Interest to help classify targets. The tilt-rotor aircrafts cannot land just before the entrance, so sending hexacopters with a camera and a lidar will allow us to determine what the entrance to the tunnel looks like and map its several initial dozen meters.

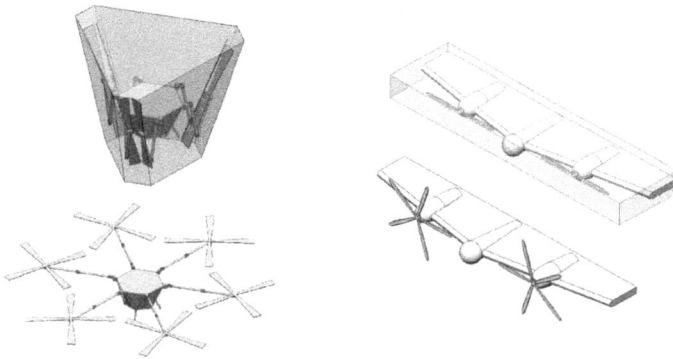

Figure 8 Rotorcraft and tilt-rotor aircraft stowed and deployed

Rotorcraft

Research shows significant potential for scientific operations on Mars using a hexacopter of about 30 kg, housed in a 2.5 m diameter aerodynamic envelope, which can lift a 5 kg payload for 10 min of flight or a distance of 5 km. A rotorcraft allows for conducting detailed observation thanks to its hovering ability.

Tilt-rotor aircraft

The proposed UAV design belongs to the tailless type aerial system. The propulsion system of the UAV consists of two five-bladed counter-rotating propellers, each driven by a single brushless electric motor in offset nacelles. The aircraft is assumed to take off from an upright position; depending on current requirements during a given phase of flight, it may remain upright, in "hover," or change orientation to horizontal, transforming into an aircraft. The wing's centre section is self-supporting, and the wingtips can be separated, leaving most of the control surfaces necessary for hover-mode flight control. For example, this particular configuration is used in transport missions - when the wingtips are detached, the drone loses the ability to fly horizontally, simplifying control, and the centre of mass is shifted backwards. The horizontal flight capability is functional when you want to scout a large or remote area that you need to get to the mission site fast enough not to lose power while flying to your destination. This capability comes in handy when making orthophotos.

System architecture

Each of the proposed aircraft designs is unstable (every rotorcraft is unstable by design in the case of a helicopter, and the flying wing concept is causing instability in the case of fixed-wing aircraft) and needs to be controlled with a closed-loop feedback system. To provide sufficient and reliable measurements necessary to control the aircraft and meet the criteria described in High-Level requirements and risk assessment, a conception of a comprehensive flight control system (FCS) was developed. An identical system has been designed for rotorcraft and fixed-wing aircraft since they require the same data set to reduce development time and costs. FCS consists of the following elements:

Flight computer (FC) - it's the main component of aircraft avionics. It receives signals from all sensors and assesses them with the Fault Detection, Isolation, and Recovery System (FDIR), which checks data integrity and accuracy and, in case of a detected sensor fault, reconfigures the system to acquire correct data from redundant sensors or estimates of inaccurate measurements. All control system algorithms run on FC, and steering inputs are sent from FC to actuators (control surfaces, motors).

Mars Integrated Guidance and Navigation Unit (MIGNU) - is designed as Linear Replacement Unit, which means it could be easily unplugged and replaced without requiring system reconfiguration. MIGNU contains all necessary sensors required to control the aircraft and perform basic scientific measurements. The aircraft's avionics system is equipped with two MIGNUs, primary and secondary, to provide redundancy and high-level fault tolerance, guaranteed by internal data validity checking and self-testing capability as well as by external assessments performed by FC.

The following sensors are integrated into MIGNU:
- The attitude measurement section is composed of an inclinometer, gyroscope, and accelerometers to determine the aircraft's orientation. Reading of angular speeds and linear accelerations are combined by an extended Kalman filter to obtain the aircraft's position in the Mars reference frame (similar to the Earth reference frame)
- Air Data sensors include static and dynamic pressure sensors to measure altitude and airspeed, hygrometer, outside temperature sensors, and gas sensors.

- The obstacle sensing unit consists of the LIDAR altimeter to provide the exact altitude above ground level (as the pressure sensor provides barometric altitude readings, which are subject to errors due to differences between the pressure-to-altitude conversion model and actual atmosphere parameters) and 360-degree lidars to detect the obstacles in the horizontal plane.
- Machine vision unit. Two cameras are used to increase guidance accuracy. The Dome nav camera performs a dual function. During day flights, it tracks the Sun and determines an aircraft's yaw angle (course). During the night, when aircraft are staying on the ground, it tracks stars to determine the aircraft's geographical position on Mars, which is used to reset Inertial Navigation algorithms as they are integrating sensors measurements (also measurement errors) and thus need to be regularly updated with precise initial conditions. Orthophoto cam is used in the surveillance of exciting areas and combined with the Interial Navigation System, and it further increases its accuracy.

Sensor fusion is vital in providing reliable and accurate data used in control systems.

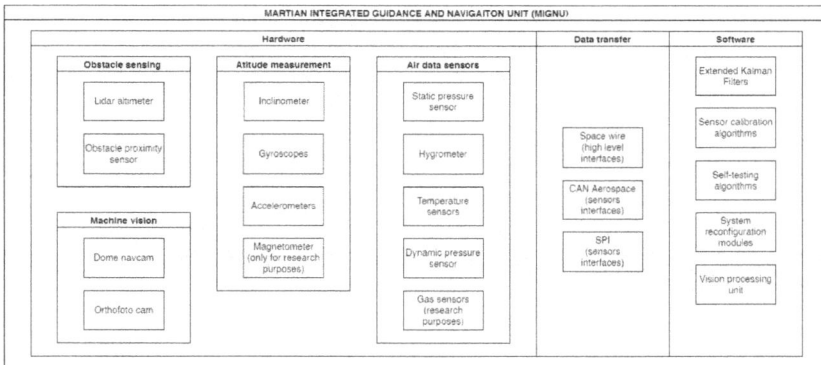

Figure 9 Mars Integrated Guidance and Navigation Unit (MIGNU)

Power system

The drones' primary power source is solar panels. By locating them in the vertical plane in the case of an aeroplane (after landing) and under the propellers - in the case of rotorcraft - the problem of dust covering the panels is eliminated. Wireless charging by a rover sent from the base can be an emergency option.

Scientific instruments

As one of the functions of flying vehicles is conducting research, in addition to the abovementioned sensors, Ground-penetrating radar is attached to them. Ground-penetrating radar (GPR) is a non-destructive geophysical method of surveying media such as soil, rock, fresh water, and ice that uses radar pulses to image the subsurface. In short, it's our eyes through the surface of the ground. A drone-mounted GPR can perform surveys in a fast manner with a high degree of automation by following pre-planned missions while also being usable in inaccessible areas where land vehicles (martian rovers) might struggle on difficult terrain. For geological surveys and bathymetry, the recommended speed is 2 m/s. For underground infrastructure mapping – recommended speed is 1 m/s. Depending on materials, space-tested GPR (as of RIMFAX) images the subsurface stratigraphy to more than 10 meters in-depth, with vertical resolutions better than 30 cm and a horizontal sampling distance of 10 cm along the rover track.

In the case of our mission, drone-mounted GPR is used to find lava tubes and possible entrances quickly. This dramatically speeds up the selection process of a lava tube to be explored, as obtaining such data directly from the near surface can even tell us how long a particular lava tube or its branch is.

Minibots

Despite using different types of robots to explore lava tubes, we decided to expand our platoons with other flying robots that can explore very narrow or hard-to-reach lava tubes. Minibots are ball-shaped robots with a diameter of approx. 20 cm. They only consist of a few components. They include the housing and frame of the robot, and inside there are electronics in the form of a control computer, batteries, elements for communication with the robot transporter and a stereo camera with an illuminator, as well as a miniature lidar. The main drive is responsible for the propulsion, and the robot's position during the flight is corrected with a set of small RCS. The camera and lidar study the surroundings and create 3D maps. This allows you to assess what is in hard-to-reach places and decide whether in the future it is worth preparing a dedicated scientific mission to this place or whether it is a place worth exploring by a human during a future crewed mission.

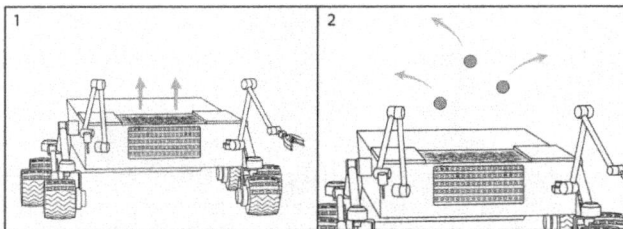

Figure 10 Minibots in transporter rover

MISSION STAGES

Mission stage I (Landing and reconnaissance)

After landing, the operation of the systems in the hold is checked. These systems are responsible for maintaining vehicles during transport (cooling and power supply) and properly unpacking after landing. After confirming the operation of all elements, the robots leave the hold. Tilt-rotor aircrafts and hexacopters leave on unique disposable platforms. After the landing, all the devices must be checked and calibrated again to adapt to Martian conditions. With calibration targets, the camera adjusts itself to show the correct colours.

Meanwhile, the satellite passing over the landing site determines the exact landing site of the lander. On this basis, scientists set the closest targets to check in the form of promising tunnel entrances. The first recognition is made using tilt-rotor aircrafts due to the extensive range and speed. The area near the entrance is mapped from the air and the entrance itself. On this basis, it is determined whether a given entrance to the tunnel is located in an area where the robot transporter can move and the route to it does not have places where the transporter rover could get stuck. Scouting with tilt-rotor aircrafts takes place, lands before entering the tunnel and scans the first few dozen meters of the tunnel. If the tunnel is clear and there are no obstacles in the fragment we have examined, the journey to the tunnel takes place using a robot transporter. This journey to speed up the entire mission begins with the tilt-rotor aircraft take off due to the differences in speed between flying and riding vehicles.

Mission stage II (Lava tubes exploration)

After reaching the tunnel entrance, the transporter rover enters an unexplored tunnel. Then the spots are unloaded, and scanning of the tunnel begins. They create a map of the tunnel and photograph the tunnel. Then the Spots return to the transporter rover, where this data is processed, based on which the following steps are determined. If the Spots come across a place they cannot explore, Minibots are sent there. If the Spots come across a place of geological interest or one where we could look for life, a decision is made to investigate this place more thoroughly with the transporter rover. If nothing interesting is found, the Spots are loaded, the transporter rover moves on, and the whole process is repeated.

| (1) TRANSPORTER ROBOT | (2) SPOTS | (3) MINIBOTS |

Figure 11 Cut-off of the Lava Tube

Mission stage III (Samples collection)

The transporter rover is equipped with a robotic arm, at the end of which is a head with a sampling mechanism. Inside the robot is a system for collecting and packing samples into transport containers. The containers are then intercepted by the walking robots that deliver them to the exit of the Lava Tubes.

The sampling process is complex and has several stages. The first stage is selecting a potential place, during which we follow the following objectives:

- Mars geology, learning about the geological processes and the history of the creation of mars and lava tubes, mainly searching for water courses, learning about the evolution of the geology of Mars.
- Searching for life, discovering the biological history of Mars, along with the return of the sample with potential life back to Earth (the chemistry of carbon and bio-signatures).
- The explanation of the processes that influenced the modification and formation of the mantle crust and the Martian core.
- Searching for dangers that may threaten the future human mars mission so that we can counteract them and determining the types and possibilities of using Martian resources to support a future mission.

Using cameras and lidar, the space around the robot is mapped. The sampling site can be selected in two ways, either using artificial intelligence algorithms or manually, with the help of scientists on Earth. The selected site is then examined with NIR and XRF instruments to determine elemental chemistry and mineralogy characterization and identification of water-related minerals. Measurements with this instrument are made without contact.

The robot has two robotic arms with 5 degrees of freedom, about 2.5 m long. At its end, there is a set of mechanisms for taking the sample and preparing the surface for sampling. It consists of a brush, a drill for discovering deeper layers, a drill for collecting the core sample and a drill for collecting regolith, i.e. the surface layer of Martian soil. These drills are interchangeable, as well as additional 2 spare pieces of each so that in the event of problems with the drill bit jamming in the material, the robot could detach the drill and continue the mission without any problems.

After sampling, the sample goes to the sample handling system. There is a drill hole in the front of the robot, and the system draws the collected material into a clean and sterile tube-shaped container and then seals it tightly to prevent contamination. The closed sample containers are placed on the front of the rover, from which the empty containers are also taken.

Mission stage IV (Sample return mission)

After the rotorcraft has landed at the lander, the sample is handed over using a robotic arm in the laboratory placed in our lander. The sample is then tested using the Raman Laser Spectrometer, XRD, MOMA and gas chromatography instruments. We collect and transport several samples from each place. Thanks to this, we can send one of the samples to Earth in the case of exciting test results. After the sample is qualified for sending to Earth, they are transported to the return rocket after collecting the appropriate loads. Then the rocket leaves the hold in a special cart and moves away to a safe distance from the laboratory, where the rocket launches. The rocket enters orbit, where it waits to be picked up by spacecraft, transporting it to the ground.

Mission stage V (Future exploration)

In addition to the research mission and the search for components necessary for future crewed exploration, the robots have been prepared to assist in future crewed missions through a modular construction method. This concept assumes that after searching the Lava Tubes, the robots return to the lander, waiting for a new task. Before the crewed mission arrives, we have detailed 3D maps of the surface and the canyon, allowing us to choose the best place to land. Heading towards the Lava Tubes, the rover looks for water in the form of ice found in the regolith. In addition, we examine the Lava Tubes, we know their geological structure as well as their shape thanks to the creation of accurate 3D maps. One of the concepts of Martian bases includes creating such a base in Lava Tubes, which ensure stable temperature conditions and protection against radiation. Knowing the location of the Lava Tubes entrances' location, shape, and construction allows us to design such a base. We also learn the composition of the surface in that area, which allows us to determine what materials we can create In-Situ. After completing the mission, the robots are so universal that they can begin preparations for the construction of the base without significant changes and will be able to help people during their stay on Mars. Then, astronauts can replace the modules with those that help build the base, e.g. arms with 3D printers, regolith-collecting excavators, or transform the robot into a material transport vehicle. Rotorcraft can transport light items, e.g. tools, between, e.g. a lander and a base.

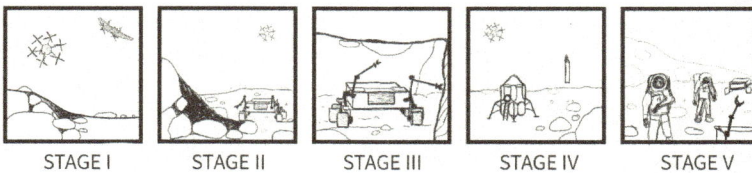

STAGE I   STAGE II   STAGE III   STAGE IV   STAGE V

Figure 12 Stages summary

MARS LANDER

Payload's weight and placing

One transporter rover with Spots and Minibots weighs 2000 kg. The rotorcraft weighs 30 kg, and the Tilt-rotor aircraft weighs 80 kg. In total, all 4 platoons with flying vehicles weigh 8,440 kg. The mass of the return rocket with the rack weighs 550 kg. The laboratory in the lander, together with the power supply and all the necessary elements, weighs 890 kg. In total, the entire mission weighs 9,930 kg. We modelled the transported cargo to check that the weight and dimensions of the proposed robots and instruments did not exceed the required dimensions.

Scientific laboratory

We decided to place some of the scientific instruments in a specially dedicated laboratory located in the lander. This allows us to use more extensive and more accurate instruments while reducing the weight of the robots. The number of instruments we can put on the transporter rover is limited by dimensions and weight, as well as the power supply. We do not test so many samples that it would be necessary to have 4 of the same sets, so we decided to create a dedicated space in the lander where the delivered samples are tested by:

- Raman Laser Spectrometer - measurements of solid powdered probes to obtain geological and mineralogical data and identification of biosignatures.
- XRD - used to validate data obtained by Raman Laser Spectrometer.
- MOMA + gas chromatograph - identification of organic molecules in collected solid samples with a mass spectrometer. Gas chromatograph - separation of complex mixtures of organic compounds into molecular components. The device contains a pyrolysis oven to preprocess the sample.

Raman Laser Spectrometer analyses the vibrational modes of a substance either in the solid, liquid or gas state. It collects and analyzes the scattered light emitted by a laser on a crushed Mars rock sample; the spectrum observed (the number of peaks, position and relative intensities) is determined by the molecular structure and composition of a compound, enabling the identification and characterization of the compounds in the sample. RLS measurements are conducted on the resulting crushed sample powder. It is a helpful tool for flagging the presence of organic molecules for further biomarker search by the MOMA analyzer. It allows us to identify organic compounds and search for life, identify mineral products and indicators of biologic activity, characterize mineral phases produced by water-related processes, characterize igneous minerals and their alteration products and characterize the water/geochemical environment as a function of depth in the shallow subsurface.

Powder diffraction of X-rays would have determined the composition of crystalline minerals. The XRD instrument also includes an X-ray fluorescence capability to provide helpful information about atomic composition. X-ray source, producing a ~ 70 μm X-ray beam that impinges on a sample. This way, we can identify the concentrations of carbonates, sulphides or other aqueous minerals to examine the past Martian environmental conditions and establish the accurate mineralogical composition of soils and rocks.

MOMA first volatilize solid organic compounds so that a mass spectrometer can analyze them; two different techniques achieve this volatilization of organic material: laser desorption and thermal volatilization (in the oven), followed by separation using four GC-MS columns (gas chromatography-mass spectrometer). The identification of the organic molecules is then performed with an ion trap mass spectrometer. MOMA helps us to find molecular biosignatures and enables further analysis of the Raman Laser Spectrometer.

Sample return

Returning the samples to Earth allows us to test them in a laboratory on Earth. The assumption of our return vehicle is a 30 kg vehicle, which can carry 5 kg of cargo into the orbit of Mars. The entire rocket, including its propellant, weighs approximately 300 kg. In addition to the rocket, a structure is created that supports it and moves it away from the lander during take-off. The rocket is two-stage, with both solid fuel engines. It is admittedly a greater mass cost than a liquid fuel rocket obtained on Mars. Still, we gain significantly in reliability and simplicity, and a heavy rocket fuel generator is also unnecessary.

COST AND SCHEDULE

If we want to send out a few platoons of robots in the coming years, we need to use technology at a high maturity level. In our project, we relied on advanced and tested solutions and used scientific instruments that had flown in previous robotic missions to Mars. As many as 3/4 of the technologies have TRL 6 or higher, as shown in the table below. R&D activities consume the most time and resources in space projects, so we can implement the assumptions of the competition to send our mission within the next decade. We also estimated how many working days (i.e. single working days for one person) are necessary to complete this task. According to estimates, we need 1,500-2 000 people working on the mission for 10 years. With an increase in the number of people, we can shorten this time, but according to our calculations, 6 years is the minimum to refine all the elements. For comparison, 3,000 NASA employees and 4,000 external specialists worked on the Curiosity rover. The proposed approach makes the mission affordable and achievable in the coming years.

L	Elements	Mission part	Previous use/Current use	TI	What we should do to use this technology in our project	Estimated time [working days]	Estimated development cost [min]	Estimated manufacturing cost [min]	Uh	Total co
1	Rover suspension	Rover	Curiosity, Perseverance	9	Rocker-bogie suspension was succesfully used during previous Mars missions. We need to scale the suspension to our needs.	62455	124,9	35,4	4	266,5
2	Rover Body	Rover	Curiosity, Perseverance	9	Rover body was succesfully used during previous Mars missions. We need to scale the body to our needs.	72378	144,8	30,9	4	268,4
3	Rover flap	Rover	Used on Earth	6	Rover flap needs proper sealing and possibility for emergency opening by spots in case of linear actuators failure	75505	151,0	30,9	4	274,6
4	Rover Cameras	Rover	Curiosity, Perseverance	9	Rover cameras were succesfully used during previous Mars missions	97766	195,5	37,9	12	650,3
5	Rover Lidar	Rover	Curiosity, Perseverance	9	Lidars were previously used on Mars landers. We need to adapt it to our needs	97766	195,5	34,5	4	333,5
6	Rover Arm	Rover	Curiosity, Perseverance	9	Rover arm was succesfully used during previous Mars missions. We need to scale the arm to our needs.	90220	180,4	37,7	8	482,0
7	Rover Drilling system	Rover	Curiosity, Perseverance	9	Drilling systems are currently used in Curiosity and Perseverance rovers	89412	178,8	28,1	4	291,2
8	Rover Sample handling	Rover	Curiosity, Perseverance	9	Sample tubes were tested on Mars missions. Sampling and sample handling were also tested during Mars missions. We need to adapt it to our needs	96861	193,7	35	4	333,7
9	Rover power generation/kilopower	Rover	Testy	9	Kilopowers were succesfully tested on Earth. Kilopower is currently being developed for future lunar and Mars missions. We need to adapt the sizes and structure of kilopower to our needs	95903	191,8	22	4	279,8
10	Rover Thermographic camera	Rover	-	6	IR cameras are used in outer-space missions. Choosen Earth-based model has to be adapted to Mars conditions	66411	132,8	33,4	4	266,4
11	Rover cooling/heating system	Rover	Curiosity, Perseverance	9	Rover radiators was succesfully used during previous Mars missions. We need to scale and adapt the radiators to our needs.	95732	191,5	29,8	4	310,7
12	Rover electronics	Rover	Curiosity, Perseverance	9	Components used on previous Mars missions. Require adaptation to sprecific mission requirements.	82962	165,9	20,5	4	247,9
13	Rover software	Rover	Curiosity, Perseverance	9	Components used on previous Mars missions. Require adaptation to sprecific mission requirements.	88472	176,9	31,5	1	208,4
14	Rover comunication	Rover	Curiosity, Perseverance	9	Components used on previous Mars missions. Require adaptation to sprecific mission requirements.	85440	170,9	26,5	4	276,9
15	Rover -> Spot recharging	Spot	NASA tests. Astrorobotic's CubeRover	6	NASA initiative developing wireless charging for lunar robots (therobotreport.com) https://spacewatch.global/2022/06/astrobotics-wireless-charging-system-survives-lunar-night/ We need to adapt sizes for Spots	71384	142,8	28	24	814,8
16	Rover -> Spot electromagnetic connection	Spot	-	6	Electromagnets. Sample magnets were used on Spirit and Opportunity https://mars.nasa.gov/mer/mission/instruments/magnet-array/	78436	156,9	34,5	24	984,9
17	Spot Thermographic camera	Spot	-	6	IR cameras are used in outer-space missions. Choosen Earth-based model has to be adapted to Mars conditions	97850	195,7	31,1	24	942,1
18	Spot lidar	Spot	Curiosity, Perseverance	6	Sample lidars were used on Mars missions. Component require certification to cosmic environment.	91279	182,6	39,9	24	1140,2
19	Spot walking system	Spot	Used on Earth	6	Joint sealing needs to be improved to withstand martian dust and weather	91537	183,1	32,2	24	955,9
20	Spot cooling/heating system	Spot	Curiosity, Perseverance	6	Additional cooling system / radiators	66842	133,7	23,9	24	707,3
21	Spot battery pack	Spot	Used on Earth	6	Additional cooling system / radiators	94180	188,4	28	24	860,4
22	Spot arm system	Spot	Used on Earth	6	is currently used by companies on Earth. We need to add some sealing protecting from Martian dust and weather conditions	68998	138,0	26,4	8	349,2
23	Spot sample transport system	Spot	Used on Earth	6	Only software developement is required, the arm can be used for sample transport	81580	163,2	31,4	8	414,4
24	Minibots structure	Minibots	University of Arizona	2/3		85068	170,1	5	108	710,1
25	Minibots thruster	Minibots	University of Arizona	2/3	Minibots have the lowest TRL in our system, however they aren't crucial part of our system and in single "wataha" we use a lot of them so failure a few is acceptable.	71233	142,5	4,8	108	660,9
26	Minibots electronics	Minibots	University of Arizona	2/3		82535	165,1	2,5	108	435,1
27	Minibots sensors	Minibots	University of Arizona	2/3	Overall the bots are very small with limited numbers of subsystem that should be developed. Part of them like sensors, electornics or battery need to be scaled to bot size.	80786	161,6	2,8	108	464,0
28	Minibots battery	Minibots	University of Arizona	2/3		96199	192,4	2,8	108	494,8
29	Minibots communication	Minibots	University of Arizona	2/3		63278	126,6	1,3	108	267,0

Figure 13 Cost of the mission

#	Item	Category	Heritage		Description					
30	SRAV first stage	Sample return	NASA MAV/ Lockheed Ma	9	Rocket stages was successfully used in earth and Mars mission, but it shuld be prepare to launch/landing loads and start form the Martian surface	71498	143.0	29.9	1	172.9
31	SRAV second stage	Sample return	NASA MAV/ Lockheed Ma	9	Rocket stages was succesfully used in earth and Mars mission, but it shuld be prepare to launch/landing loads and start form the Martian surface	85078	170.2	36.7	1	206.9
32	SRAV sample container	Sample return	NASA MAV/ Lockheed Ma	6	Sample contener should be developed and qualified	65694	131.4	36.2	1	167.6
33	SRAV sample allocation	Sample return	NASA MAV/ Lockheed Ma	6	Sample allocation system eg. Robotic arm should be developed and qualified	67553	135.1	27.6	1	162.7
34	SRAV Launch Tube	Sample return	-	2	Trolley should be developed and qualified	80322	160.6	30.6	1	191.2
35	Rotorocraft Structure	Rotorcraft	-	6	Multirotor concept is proven and viable design as shown by many Earth based aircrafts.	84642	169.3	32.7	1	202.0
36	Rotorcraft Folding system	Rotorcraft	Pathfinder	6	Folding systems were used during numerous Mars missions (e.g. Pathfinder)	82871	165.7	20.9	4	249.3
37	Rotorocraft navigation system	Rotorcraft	Ingenuity	6	All components in less advanced version were used in Ingenuity UAV.	68610	137.2	23.5	4	231.2
38	Rotorcraft flight system	Rotorcraft	Ingenuity	6	All components in less advanced version were used in Ingenuity UAV.	88200	176.4	32.3	4	305.6
39	Rotorcraft lidar	Rotorcraft	Ingenuity	6	All components in less advanced version were used in Ingenuity UAV.	70880	141.8	31.7	4	268.6
40	Rotorocraft sample transport system	Rotorcraft	-	2	Untested idea, however used in Earth conditions	77960	155.9	36.4	4	301.5
41	Rotorcraft battery	Rotorcraft	Ingenuity	6	All components in less advanced version were used in Ingenuity UAV.	51855	103.7	33.6	4	238.1
42	Rotorocraft solar cells	Rotorcraft	Ingenuity	6	All components in less advanced version were used in Ingenuity UAV.	64843	129.7	35	4	269.7
43	UAV Structure	UAV		6	Design proven in many Earth based designs	84324	168.6	22	4	256.6
44	UAV lidar	UAV	Ingenuity	6	All components in less advanced version were used in Ingenuity UAV.	71151	142.3	26.1	4	246.7
45	UAV navigation system	UAV	Ingenuity	6	All components in less advanced version were used in Ingenuity UAV.	78276	156.6	35.3	4	297.8
46	UAV battery	UAV	Ingenuity	6	All components in less advanced version were used in Ingenuity UAV.	96617	193.2	38.2	4	346.0
47	UAV solar cells	UAV	Ingenuity	6	All components in less advanced version were used in Ingenuity UAV.	82518	165.0	24.8	4	264.2
48	XRF	Rover	Perseverance, Curiosity	9	Adapt to our rover	60033	120.1	36	4	264.1
49	NIR	Rover	Rosalind Franklin	8	Adapt to our rover	84882	169.8	32.6	4	300.2
50	Raman Laser Spectrometer	Laboratory	Rosalind Franklin, Perseve	9	Adapt to our lander laboratory	63391	126.8	36.5	4	163.3
51	XRD	Laboratory	Rosalind Franklin, Curiosity	9	Adapt to our lander laboratory	68792	137.6	32	1	169.6
52	MOMA + gas chromatograph	Laboratory	Rosalind Franklin	9	Adapt to our lander laboratory	98976	198.0	22.4	1	220.4
53	Geology scanners	Rover	Perseverance	9	RIMFAX was used during last Mars mission	53391	106.8	35.4	12	531.6
54	Mars wheeater sensors	Rover	Perseverance, Curiosity	9	During previous Mars mission this type of instruments was used for atmospheric pressure, temperature, humidity, winds, and ultraviolet radiation level mesurements.	80880	161.8	16.8	4	229.0

Total cost [bln $]	21
Number of workers	1705
Cost of one worker per day [$]	2000

THE INNSPACE TEAM

Figure 14 Visualization of the project

Innspace is a group of young scientists fascinated by space exploration. Our portfolio includes award-winning projects for bases on the Moon and Mars and space vehicles. Meet the team responsible for the project:

- Piotr Torchała as a leader, Justyna Pelc and Beata Suścicka as project support,
- Science team: Ewa Borowska, Magdalena Łabowska, Nikola Bukowiecka, Aleksandra Klassa, Bartosz Rybacki,
- Engineering team: Hubert Gross, Nikodem Drąg, Michał S łomiany, Kamil Ziółkowski, Kajetan Szostek, Milena Michalska,

3: PRECURSOR
The Ultimate Telerobotic Precursor to a Human Mission to Mars
Rollers, Rotors & Unfinished Business

Carl Greenbaum
carlggreenbaum@gmail.com

SUMMARY

The mission concept and system design for PRECURSOR provides a modest risk, integrated package for maximum science return and extensive preparation for follow-on Human missions. The principal science objectives are searches for extant biological activity and fragmentary DNA/RNA in subsurface regolith and water ice and finding shallow, subsurface water ice in harvestable quantities. The principal exploration preparation objectives are harvesting the water and demonstrating In-Situ Methalox propellant production powered by Kilopower fission reactors. Secondary science objective is characterization of regolith organics. The secondary objectives for exploration preparation are an aquaponics operational assessment in micro and Martian gravity environments and evaluation of a backup power source, a pair of 2.5 kw solar arrays. A single 4 metric ton payload will deliver two Mobile Landers based on the Chariot Lunar Rover which will transport science and expedition preparation experiments to final deployment sites and four Ingenuity derived Mars Science Helicopters. (Advanced ROtary-Winged roBOTs -AROWBOTs).

INTRODUCTION

In 1976, two Viking spacecraft were the first human made objects to land on Mars. The science return was enormous but the ambiguous, to some, Labelled Release Experiment results cast a decades long pall on the search for planetary life. The most convincing criticisms of a positive labelled release result were the unexpected lack of organics detected and the apparent absence of liquid water on the Martian surface. However, subsequent analysis has revealed abundant organics and the Viking 2 Lander sampling head indicated the presence of liquid water. "As the sun rose, the temperature of the sampling head plate resting on the soil increased until pausing at $273°$ K,the unique and identifying temperature of water ice liquefying. This is strong evidence that sufficient ice was in the surface material such that absorption of the heat of fusion by the ice interrupted the rise in temperature" (Ref 1).

Back in 1976 the answer to whether there was life on Mars was important but not essential to further exploration. Now, as we stand on the threshold of a crewed Mars mission, the answer to that question is far more urgent. Beyond the "Is there other life in the universe" question, planetary protection and back contamination issues drive numerous, pervasive mission requirements. This robotic PRECURSOR mission can ensure the appropriate design criteria are applied to maximize mission success. Reference 1 makes the following recommendation, "The return to Earth of a sample of Martian surface material should be deferred until the nature of any life present is determined with respect to any possible hazard to terrestrial life forms or environment." This mission can provide that hazard determination.

SCIENCE PAYLOADS

Mars Instrument for Chiral Research on Biological Environments (MICROBE): A worthy successor to the Viking Labelled Release Experiment, (Figure 1) this instrument will use the same basic operating procedure. The labeled release experiment was based on the assumption that any extant life on Mars would metabolize simple carbon compounds and produce gaseous wastes. Therefore, the experiment called for the incubation of a sample with a low concentration nutrient of radioactively marked carbon molecules and the subsequent testing for gaseous products containing the radioactive elements.

The MICROBE experiment can provide an unambiguous distinction between non-biological and biological agents. The key determinant is its ability to detect chiral preferences in the utilization of substrates applied to the sample. Strong preference for L-amino acids and D-carbohydrates is a peculiar, still unexplained, property of all known living systems. Such preferences have not been reported as naturally occurring in chemical reactions. The original Viking experiment included nutrients

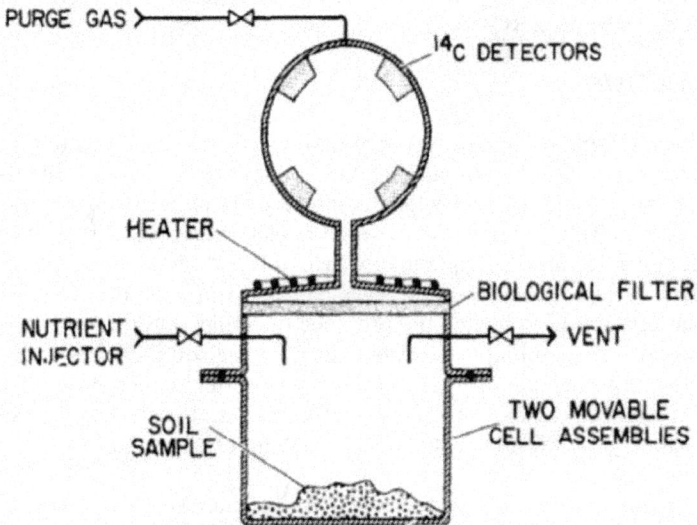

Fig. 1 Labelled Release Process

of both chiralities but they were mixed together since, to isolate them would have doubled the number of nutrient samples hence the size and mass of the instrument.

Two different variants of the MICROBE experiment will be used. The Chariot rover will have an instrument configured in the rover body similar to the Viking installation (MICROBE-R). The AROWBOTs will have a carousel of single sample instruments (MICROBE-A Figure 2 Ref 2).

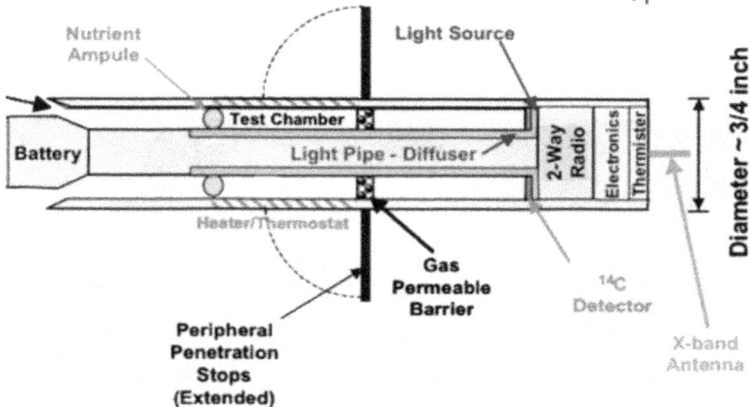

Fig. 2 Labelled Release Single Sample Device

SIngle Molecule DNA Sequencing (SIMDNA): DNA and RNA are molecules that are present in all living things on earth. DNA is the genetic code of life, the instructions for building and operating an organism. RNA is primarily a messenger molecule, carrying instructions from the DNA code to control the synthesis of proteins. While there is no guarantee that Martian life would also use these genetic molecules, a search for them certainly deserves a prominent place in the search space. "It seems improbable to me that we will do a serious search for life on Mars and not do this test." Dr. Chris McKay, NASA Ames.

The MIT/Harvard Search for Extra-Terrestrial Genomes (SETG) Project wants to test the hypothesis that life on Mars, if it exists, shares a common ancestor with life on Earth. There is increasing evidence that viable microbes could have been transferred between the two planets, based in part on calculations of meteorite trajectories and magnetization studies supporting only mild heating of meteorite cores. Based on the shared-ancestry hypothesis, this instrument will look for DNA and RNA through in-situ analysis of Martian soil and ice samples.

The SETG project is designing an instrument for just such a search based on single molecule DNA/RNA sequencing using the Oxford Nanopore Technologies' MinION (Fig. 3). Oxford Nanopore sequencing devices use flow cells which contain an array of tiny holes — nanopores — embedded in an electro-resistant membrane. Each nanopore corresponds to its own electrode connected to a channel and sensor chip, which measures the electric current that flows through the nanopore. When a molecule passes through a nanopore, the current is disrupted to produce a characteristic 'squiggle' (Fig 4). Each base that passes through the nanopore can be identified through the characteristic disruption it causes to the current in real-time.

The process has been tested in a chamber at Martian temperatures and pressure and on the Zero G aircraft simulating Martian and micro gravity and on the ISS. The prototype has performed well in these tests and work is currently underway to produce a fully automated device to gain confidence that a flight ready instrument is achievable.

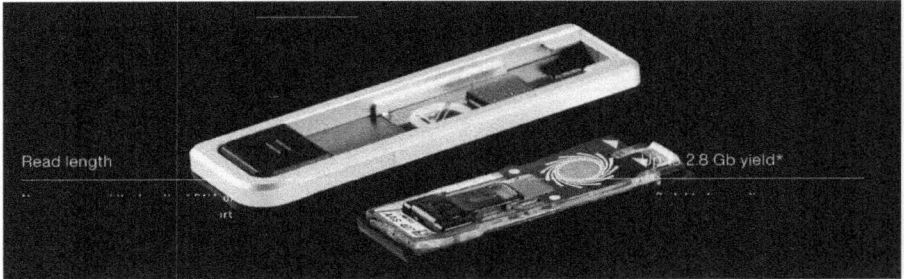

Read length   2.8 Gb yield*

Fig. 3 SIMDA Single Sample Device

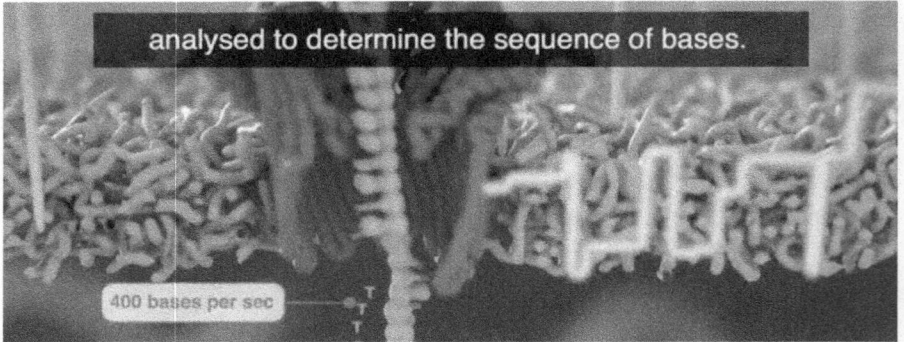

analysed to determine the sequence of bases.

400 bases per sec

Fig. 4 Nano-Pore "Waveform"

SIMDNA can detect both extant and recent DNA/RNA based biological activity. The lengths of the surviving nucleic acid strands may also provide insight into the time since the fragment was whole. Ref 3 assessed the half-life of ancient earth DNA at 521 years and concluded, "If the decay rate is accurate then we predict that DNA fragments of sufficient length will preserve in frozen material of around one million years in age"

In a manner similar to the MICROBE instruments, a multi-sample instrument will be installed on the rover (SIMDNA-R) and a carousel of single sample instruments will be carried by the AROWBOTS (SIMDNA-A). Subsurface regolith or water ice samples will be delivered to the instrument.

<u>Sample Acquisition Drill:</u> Samples for the MICROBE and SIMDNA instruments must be taken from subsurface material. Figure 5 illustrates the penetration of radiation harmful to biology and preserved DNA.

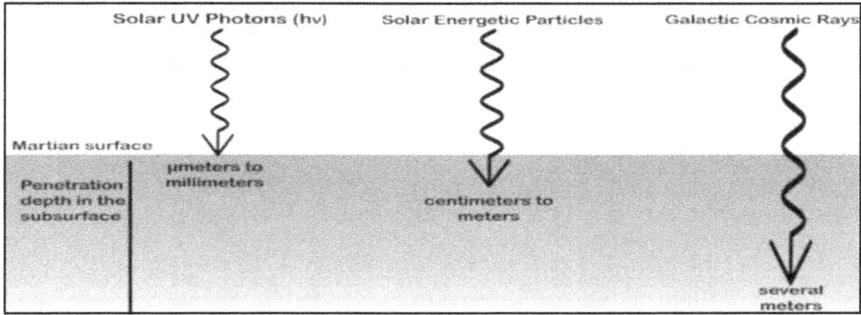

Fig. 5 Radiation Penetration on Mars

At a minimum, samples should be collected a few cm below the surface. This will ensure samples free from UV damage. However, energetic solar particles can also cause biological damage so samples 1-2 meters down are desired. For the airborne instruments a drill that collects powder samples a few cm down will have to suffice. A lightweight derivative of the Curiosity Powder Acquisition and Delivery System (PADS) can do this if the mass can get below 1 kg (Fig. 6). For rover-based instruments a deeper drill is desired, like the drill on the ExoMars rover, Rosalyn Franklin, capable of retrieving samples from up to 2 m below the surface (Fig. 7).

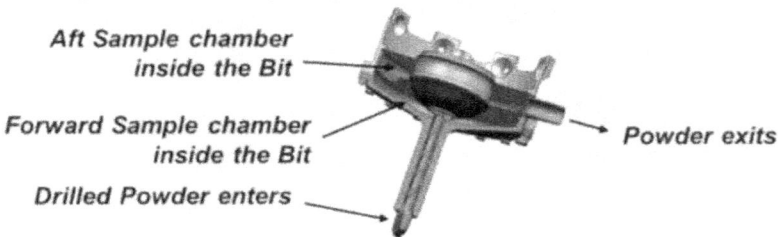

Fig. 6 Curiosity's Sample drill0

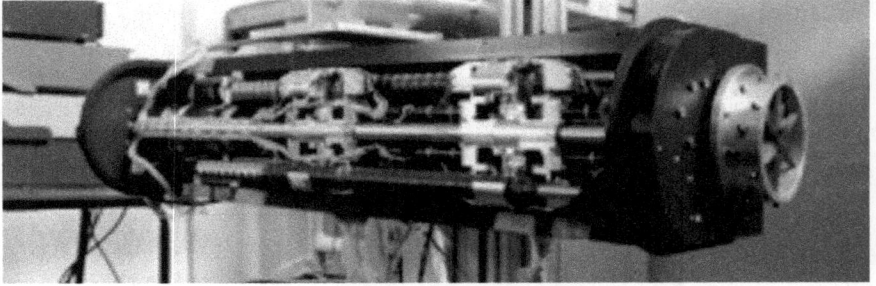

Fig. 7. ExoMars 2-Meter Drill

Radar Imager for Mars Subsurface Experiment (RIMFAX) will provide ground penetrating radar data in search of near surface water ice deposits. Since the mass is only 3 kg, it will be deployed on one AROWBOT helicopter for larger search radii than the CARTs. Data can be taken landed or at low altitude hover. Figure 8 shows the electronics module. Since the mass of the RIMFAX limits the primary science payloads, only one will be used. The other AROWBOTs will carry higher capacity MICROBE-A and SimDNA-A instruments.

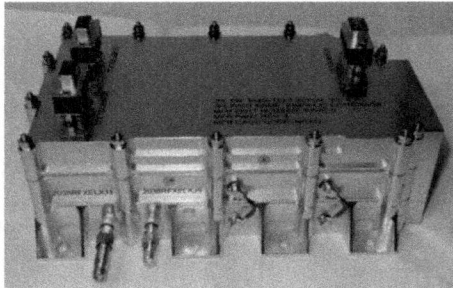

Fig. 8 RIMFAX Hardware

Scanning Habitable Environments with Raman & Luminescence for Organics & Chemicals (SHERLOC): Provides Fine-scale detection of minerals, organic molecules and potential biosignatures. Originally mounted on Perseverance's robotic arm, SHERLOC uses cameras, spectrometers, and a laser to search for organics and minerals that have been altered by watery environments and may be signs of past microbial life (Figure 9). In addition to its black-and-white context camera, SHERLOC is assisted by WATSON, a color camera for taking close-up images of rock grains and surface textures. For the PRECURSOR Mission, SHERLOC will be mounted on the Science Chariot Rover.

Fig.9 SHERLOC Hardware

EXPLORATION PREPARATION PAYLOADS

Water Harvesting: The use of terrestrial ice drilling and Rodriguez Well techniques have promise to generate a source of liquid water from presumptive Martian sub-surface ice for an operational system at Mars. However, the heat input available and water withdrawal rates for a representative Mars surface mission are small compared to most terrestrial experience. Development tests using a functional prototype of such an operational system will validate the analytical results.

Water will be harvested by drilling a Rodriquez Well (Rodwell) and melting a pocket of ice from which water will be extracted. Figure 10 illustrates the process. The power needed to melt and extract the water must be balanced with the extraction rate to avoid melting water that is not needed or pumping the well "dry" by attempting to pump water out faster than it is being melted. Reference 4 describes a "Redwater" system for Mars water harvesting. That analysis concludes that, to produce the water mass for MAV propellant, numerous smaller wells are more power efficient than one large well. For PRECURSOR, a single well will serve to demonstrate the technology. A device is described with an auger 6 cm in diameter and 10 cm long with associated components to establish and maintain the Rodwell. This device has achieved a TRL of 4/5 with testing planned to bring it to TRL 6. The Redwater prototype can drill through up to 25m of overburden using a Coiled Tube Drill that uses compressed gas to eject drill shavings. The METHALOXIE variant will be limited to 4m depth based on expected overburden. For this depth, 2 kg of gas are required. This would be a minor mass penalty but an alternative is to employ a scroll pump similar to that used on MOXIE that could compress the needed CO_2 from 7 to 760 Torr in less than 1 sol for 4 kwh of electrical energy (160 w).

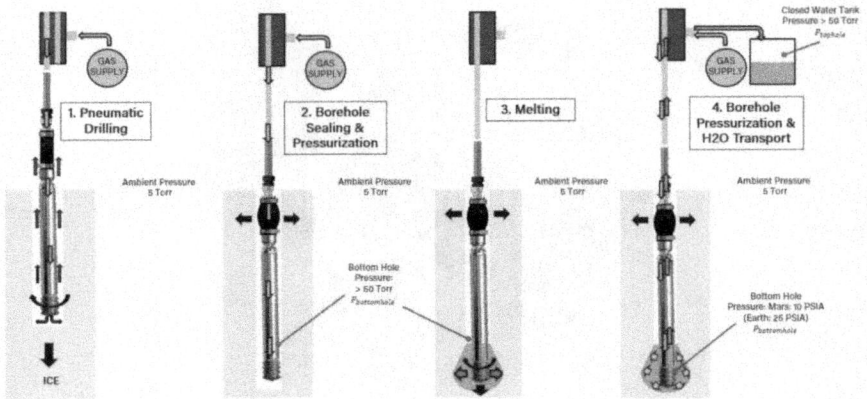

Fig. 10. Water Harvesting with Rodriquez Well

Methalox In-Situ Experiment (METHALOXIE): The best know approach to defeat the "Tyranny of the Rocket Equation" is In-Situ propellant production. Anticipated propellant needs include Liquid Hydrogen/Oxygen (Hydrolox) and Liquid Oxygen/Methane (Metalox). Mars has the necessary raw materials for both of these options. Hydrolox is "easy" in that electrolysis of water will produce H2 and O2 that can be cooled/liquified.

Methalox production requires an additional step beyond electrolysis. The Hydrogen must then be combined with CO2 from the Martian atmosphere to produce Methane and water. The technique is the well understood Sabatier Process.Power-to-fuel systems via solid-oxide electrolysis are promising for storing excess renewable electricity by efficient co-electrolysis of steam and CO2 into methane and oxygen. This rapidly developing technology will permit a mass and power efficient Mars ISRU fuel production system. Prototype systems currently in test can produce 1kg/day of oxygen and methane using 17 kwh/day (700w continuous). This system has a mass of 50 kg. A typical solid oxide electrolysis cell is shown in Figure 11.

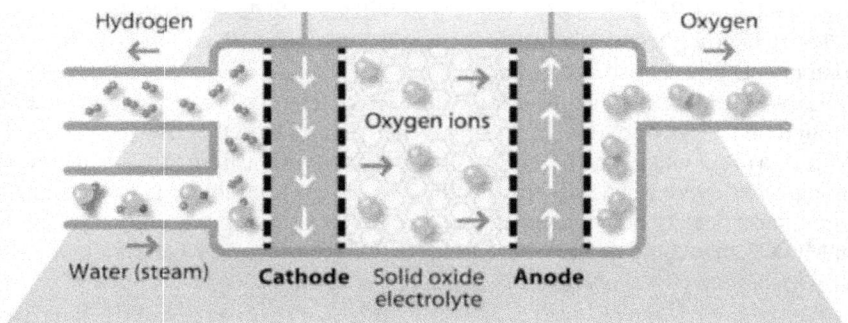

Fig. 11 SOEC Operation

The Human Mars Ascent Vehicle described in Reference 5 will have a dry mass of 5 mt and a propellant mass of 19 mt. For the 26 month Martian launch window and 7 month transit, 19 months of propellant production time is available or 1,000 kg/month.

For this PRECURSOR demonstration system, production of 10% of this propellant mass is a reasonable goal. Figure 12 illustrates the METHALOXIE operation. The oxygen and methane masses cited reflect 30 sols of Methalox production. Table 1 lists the system Parameters. Energy requirements are based on Reference 6.

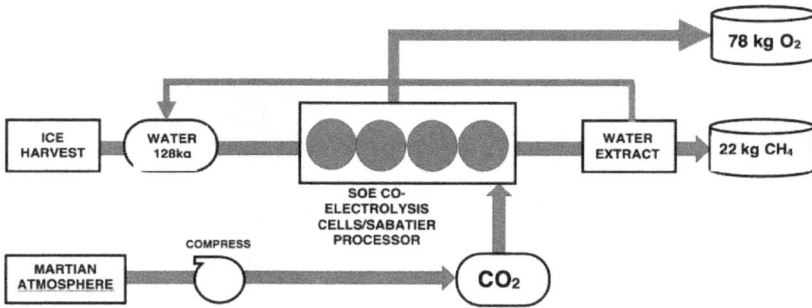

Fig. 12 METHALOXIE Process Diagram

Table 1-METHALOXIE System Parameters

COMPONENT	EMPTY MASS(kg)	FULL MASS (kg)	POWER (w)
Water Tank	15	65	75
Co-Electrolysis Stack	150	150	2100
CO2 Compressor	2	2	150
CO2 Tank	2	6	0
Methane Tank	5	27	0
Oxygen Tank	10	88	0
Water Extractor	2	2	20
Totals	**186**	**340**	**2390**

Kilopower Reactor: The Kilopower reactor project evolved into the Fission Surface Power Project in 2018. Kilopower developed and tested a 1 KW reactor. That reactor, called KRUSTY, used 25 kg of highly enriched Uranium, contained in a Beryllium Oxide cannister and moderated by a boron rod. Fission energy was delivered to Sterling converters via sodium filled heat pipes. The critical design characteristic is self regulation. Spike loads and even total load withdrawal were handled without the need for complex control circuitry (Ref 7). The design intent was to scale the KRUSTY demonstration reactor as high as 10 KW. It is estimated that Kilopower type units generate about 4 times as much thermal energy as electrical energy. Some of that thermal energy can be available for ice melting and heating the electrolysis and Sabatier processors.

However, the Kilopower project concluded in 2018 and NASA launched the Fission Surface Power Project with a goal of developing a 40 KW reactor capable of fully supporting a crew of 6 on the Martian surface. NASA is collaborating with DOE and industry to design, fabricate, and test a 40-kilowatt class fission power system to operate on the Moon by the late 2020s. So far, only Memoranda of Understanding with DOE. The project lead for the Kilopower program reported spending $20 million on KRUSTY and believed the 10 KW version could be built for $100 million. Since 2 kw will power the METHALOXIE and other experiments, the lower risk and lower cost approach is to use derivatives of two KRUSTY type reactors (Figure 13).

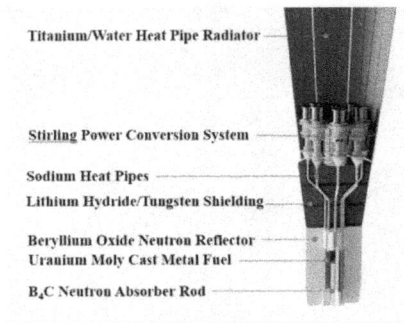

Titanium/Water Heat Pipe Radiator

Stirling Power Conversion System

Sodium Heat Pipes

Lithium Hydride/Tungsten Shielding

Beryllium Oxide Neutron Reflector
Uranium Moly Cast Metal Fuel

B₄C Neutron Absorber Rod

Fig. 13 1kw KRUSTY Kilopower Flight Concept Reactor

Solar Panel Option: In spite of the success of the KRUSTY program, there are non-trivial technical and regulatory risks associated with fabrication and deployment of space qualified 1kw Kilopower reactors. Also, kw/kg of photovoltaic solutions are rapidly improving. Parallel development and deployment of a solar array solution is, therefore, included. This system will be based on the Roll Out Solar Array (ROSA) designed as a replacement for ISS panels. The first two have already been installed. The stowed size is 0.6m Diameter x 3m long (Fig 14). When fully deployed, the half length and width array is 3m x 10m and produces 2.5 kw in full sun at Mars. The array has a mass of 85 kg. A battery system is required to support operations during the Martian night. Assuming operation around the equinoxes, battery backup will be required for 12 hours and 20 minutes per sol, roughly 25 kwh of battery backup. A second 2.5 Kw ROSA array will be required to recharge the batteries during daylight. Advanced Lithium-Ion batteries are projected to soon achieve energy densities of over 650 wh/kg (Ref 8). This design assumes 450 wh/kg. The mass of the resulting battery is 58kg. Combined with the ROSA mass of 85kg, a complete system of two arrays with charging and control electronics will be approximately 255 kg.

Fig.14. Stowed Roll-Up Solar Array

Mars Experimental Aquaponics Laboratory System (MEALS): Nutritious food is essential for long duration planetary exploration. Aquaponics (integrated production of plants and fish) holds promise as a continual source of fruits, vegetables and fish protein (Ref 9).

However, startup delays until plants mature and fish grow to harvest size are considerable. Starting from scratch after landing, the grow out phase could consume more than half of the surface stay of a conjunction class Mars mission. An effective system needs a head start.

The objective of this experiment is to demonstrate an autonomous micro-scale aquaponics system that can survive the launch shock and vibration environment, operate effectively during coast phase in micro-gravity and arrive on the Martian surface with a small crop of nearly mature fruits and vegetables and a small population of fish nearly ready to breed and/or harvest. Tables 2 and 3 list candidate fish and plants.

Table 2 Candidate Fish Species

FISH	TEMP RANGE	GROWTH RATE
Tilapia	18-30	Fast
Trout	14-20	Slow
Catfish	26-30	Moderate
Bass	24-30	Moderate
Salmon	13-18	Slow
Prawns	24-30	Fast

Table 3 Candidate Plant Species

ITEM	TEMP RANGE	pH RANGE
Lettuces	Cool	6.0-7.0
Tomatoes	Hot	5.5-6.5
Radishes	Cool	6.0-7.0
Kale	Cool-Warm	5.5-6.5
Cucumbers	Hot	5.5-6.0
Spinaches	Cool-Warm	6.0-7.0
Beans	Warm	6.0
Chives	Warm-Hot	6.0
Basil	Warm	5.5-6.5

The basic nutrient flow of an aquaponics system is shown in Figure 15.

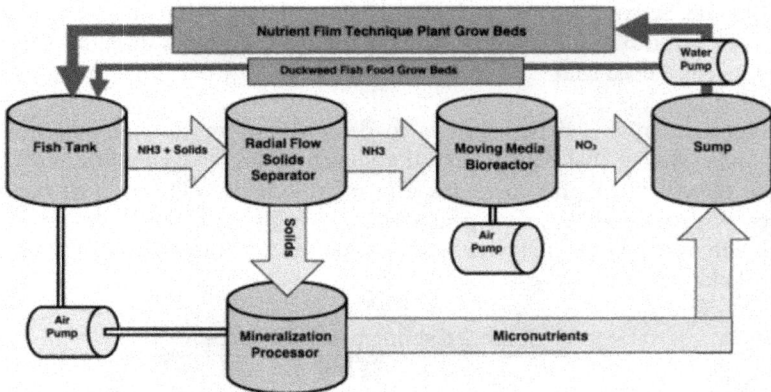

Fig. 15. Aquaponics Nutrient Flow

To accommodate micro gravity operations, free water surfaces are avoided. The fish tank and filter volumes run completely full and an integrated oxygen bottle keeps the dissolved oxygen level at desired levels. Feeding, heating and pH control are automatically controlled. Starter plants in the Nutrient Flow Tubes (NFT) are sealed in to prevent leakage. The fish population at launch is sized to allow 8 months of growth to reach tank holding capacity. Fish and plant development are monitored by video link and system operational parameters are downlinked as well. The automated control system is shown in Figure 16.

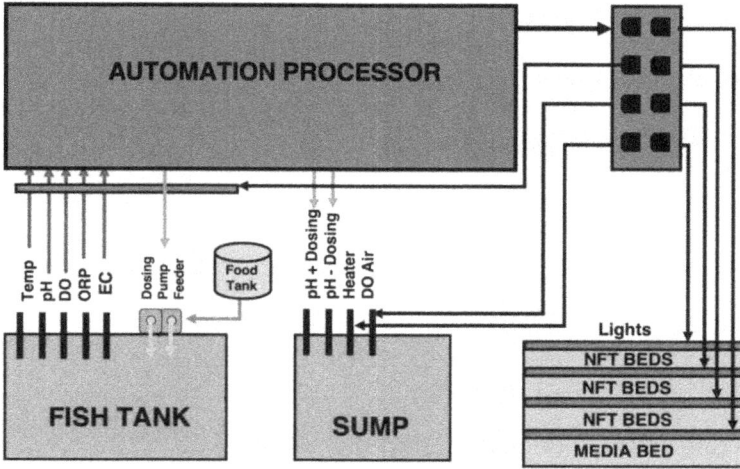

Fig. 16 MEALS Automated Control System

The layout of tanks, filters and grow beds is shown in Figure 17. The tank section is similar to the Aquarium Habitat (AQH) used on the ISS. Multiple ISS experiments using a variety of fish species have demonstrated their ability to survive launch shock and vibration, orient themselves to illumination sources in microgravity and to breed successfully (Ref 10).

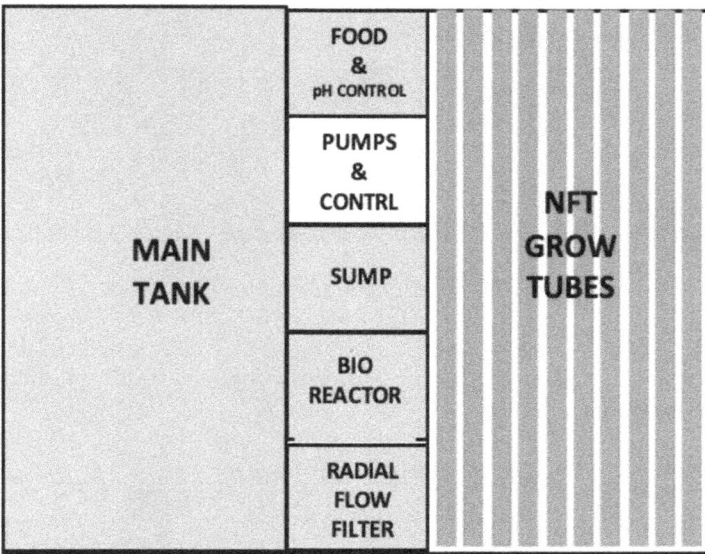

Fig. 17 MEALS Components

ROBOTIC SYSTEM DESIGN

Chariot <u>Autonomous Rover Transporter (CART)</u>: This rover is an evolution of the earlier Lunar Electric Rover "Chariot" 12 wheeled chassis (Fig. 18, Ref 11). Specifications are summarized in Table 4.

Table 4 CART Specifications

Mass (KG)	1,000
Payload (KG)	3-6,000
Length (M)	4.5
Height (M)	1
Speed (KPH)	20
Range (KM)	100

Fig. 18 Chariot Autonomous Rover-Transporter

The PRECURSOR Mission will carry two CARTs. One dedicated to science objectives and a second supporting ISRU demonstrations. Table 5 lists the different CART payloads.

Table 5 CART Payload

SCIENCE CHARIOT	MASS (KG)	ISRU CHARIOT	MASS (KG)
MICROBE-R	50	Rodwell Water Extractor	100
SIMDNA-R	40	METHALOXIE	186
SHERLOC	4	MEALS Demo	360

ExoMars 2 M Drill	25		0
1 kw Kilopower	400	1 kw Kilopower	400
Roll-Out Solar Panels	170	Roll-Out Solar Panels	85
AROWBOTs (4)	124		
Total Mass	813	Total Mass	1,131

Advanced Rotary Wing Robot (AROWBOT): The airborne element of this robotic system is a derivative of the wildly successful Ingenuity helicopter delivered to Mars by the Mars 2020 Rover Perseverance. While classified as a Technology Demonstration, Ingenuity has flown many more than the planned five flights. Understandably, the design, other that the flight system, was uncomplicated with minimal payload.

Even before Ingenuity arrived at Mars, JPL engineers were studying concepts for a more capable Mars Science Helicopters (MSH). Reference 12 analyzes a range of payload/mission duration options. For our robotic system, the primary roles of the airborne elements are to search for near surface water ice and sample subsurface regolith and ice with the MICROBE and SIMDNA instruments. Based on these candidate payloads, an MSH variant with 5 kg science payload and 10 km range was selected for this mission. Table 6 compares the AROWBOT parameters to Ingenuity. The AROWBOT will incorporate a small diameter drill/auger to deliver subsurface regolith and water samples to the MICROBE and SIMDNA instruments and to calibrate the RIMFAX data. The MSH is illustrated in figure 19. AROWBOT Payloads are listed in Table 7.

Table 6 Mars Helicopter Comparison

PARAMETER	INGENUITY	AROWBOT
Mass (kg)	1.8	31
Science Payload (kg)	0.2	5
Rotor Diameter (M)	1.2	1.2
Number of Rotors	2	12
Blade Speed (RPM)	2575	2943
Battery (AH)	12	46
Solar Panel (cm^2)	400	600
Fuselage Size (cm)	14 x 20 x 16	20 x 30 x 20
Camera	Color 4208 x 3120 px	Color 8,000 x 6,000
Range (km) or Hover Time (Sec)	NA/30	10/300
Speed (m/s)	2	30
Instrumentation	None	MICROBE, RIMFAX, SIMDNA

Table 7 AROWBOT Payloads

COMPONENT	ARROWBOT-1&2 MASS (KG)	ARROWBOT-3 & 4 MASS (KG)
MICROBE-A	0.5	2.0
SIMDNA-A	0.5	2.0
RIMFAX	3.0	0.0
PADS	1.0	1.0
Total	5.0	5.0

Fig. 19 AROWBOT Variant of Mars Science Helicopter

An Alternative System Configuration: It may turn out that the AROWBOT mass limitation imposes unsatisfactory design constraints on the drill or other payload elements. An alternative configuration is to limit the AROWBOT payload to sample collection and design the MICROBE-R and SimDNA-R payloads to accept and process samples delivered by the AROWBOTS. Detailed design trade-offs are appropriate.

CONCEPT OF OPERATIONS

Landing Site: Figure 20 is a map of water ice depth below the Martian regolith. The white rectangles highlight a preferred landing sites based on ice location and quantities. The region is at latitudes 35 to 40 degrees north and an elevation of about -4,000m. The regolith overburden to the water ice is around 0.1m.

Fig. 20. Regolith Thickness over Water Ice

An alternative landing site is in the southern highlands. "The southern hemisphere has ancient rocks based on higher crater counts and crustal magnetism. The region around the Polar Explorer planned landing site (fig. 21) is a great place to search for signs of ancient life,"- Dr Chris McKay

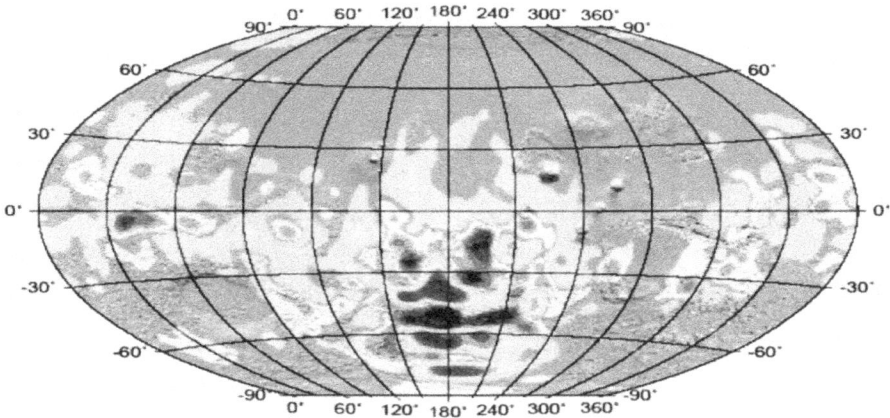

Fig. 21 - Crustal Magnetism Map

Launch Configuration: No details are available to the design teams regarding the launch vehicle integration or delivery to the Martian surface. Figure 22 illustrates a vertically integrated launch stack for all system components. It has been assumed that equipment will be craned off of the landing vehicle or driven off of ramps for the wheeled robots.

Packaging of the mission hardware within the specified 5m diameter fairing is based on the following sizing information:

- The CART Rover prototypes are 4.5m long x 3.7m wide x 1.25m high
- The MSH upon which the AROWBOTS are based, were designed to fit within a 2.5m x 1.6m aeroshell. The fourth AROWBOT is behind the top unit shown.
- The METHALOXIE tanks and processors have been designed to fit a 1m high package shaped around the CART suspension.
- The ROSA Arrays and Kilopower radiators are less than 0.6m diameter and fit between the CART wheels. A second set of each is behind the CARTS

Fig. 22 PRECURSOR Mission Launch Stack Configuration
(5m Diameter x 5m High)

Operational Sequence: Two AROWBOTs deploy, self-test and begin sampling sites with promising morphological features. The other Two AROWBOTs deploy, self-test and begin RIMFAX searches for large, shallow water ice deposits. When the best available site is found, the ISRU CART will deliver the METHALOXIE with Rodwell drill plus the connected Kilopower unit to the ice location and activate the fission reaction. Once the fission reaction achieves stable power output, the ice melting and propellant production will commence and operate for 30 sols.

The Science CART will search for promising sample sites within driving distance and commence science operations. Holes will be drilled and subsurface regolith, and possibly water ice, will be draw into the MICROBE-R and SIMDNA-R instruments for analysis. Once samples have been obtained, SHERLOC will commence operations.

Once the initial ice location has been found, the AROWBOTS will begin search for promising sites for MICROBE-A and SIMDNA-A sampling beyond the range of the Science CART. At maximum single flight range, the AROWBOTs will land to recharge during the next sol then continue searching/sampling.

MASS ESTIMATE (KG)

COMPONENT	MASS	QUAN	TOTAL	REFERENCE
Chariot Autonomous Rover	1000	2	2000	Prototype Mass
AROWBOTs	31	4	124	Upgrade from 1.8 kg
1 KW KiloPower Reactor	400	2	800	KRUSTY Estimate
ISS Roll-Up Solar Panels	85	2	170	Current flight hardware
Battery & Control Electronics	85	1	85	450 wh/kg Batteries
MICROBE-A	2	4	8	
MICROBE-R	5	1	5	
SIMDNA-A	2	4	8	
SIMDNA-R	5	1	5	
RIMFAX	3	2	6	Flight Instrument
AROWBOT Drill	1	3	3	
CART drill	15	1	15	
ISRU Propellant Plant	186	1	186	Bill of Materials Estimate
Aquaponics demonstration	360	1	360	Bill of Materials Estimate
Water Harvesting System	100	1	100	
Total			**3,875**	**Kg**

COST ESTIMATE ($M)

COMPONENT	COST	QUAN	TOTAL	REFERENCE
Chariot Rover (CART)	250	2	500	Estimated space qual of prototype
AROWBOTs	100	4	400	$80M less NRE + add'l rotors
1 KW KiloPower Reactor	40	2	80	2 X KRUSTY
ROSA, Battery & Control	46	2	92	Panel=45, Batt+control=1M
MICROBE-on Rover	100	1	100	Similar Instrument Costs
MICROBE-on AROWBOTs	2	40	80	
SHERLOC	15	1	15	
ISRU Propellant Plant	250	1	250	2 X MOXIE & tanks, comp, evap
Aquaponics demonstration	10	1	10	AQH plus automation
Microwave Ice Harvest System	10	3	30	Commercial parts x 10
Operational Support (12 months)	250	2	500	2X Mars2020 Ops Costs
Total			**2057**	**$M**

SCHEDULE ESTIMATE

The original Mars Science Laboratory (MSL) schedule from Mission Concept Review (MCR) and Announcement of Opportunity (AO) for instruments to launch is shown in Figure 23. Even though the actual MSL development took much longer, this still represents a reasonable estimate for the PRECURSOR development. Not only does it include some margin for retiring risk, it contains time for design and development of the sky crane Entry, Decent and Landing (EDL) hardware and software, a substantial schedule risk not part of this program. If the MCR for the PRECURSOR program occurs in January of 2023, this schedule will meet the 2028/2029 Mars rendezvous window. Candidate launch and arrival dates with associated durations and Delta-Vs are shown in Table 8. With a launch in January 2029, the mission will arrive at Mars in July 2029.

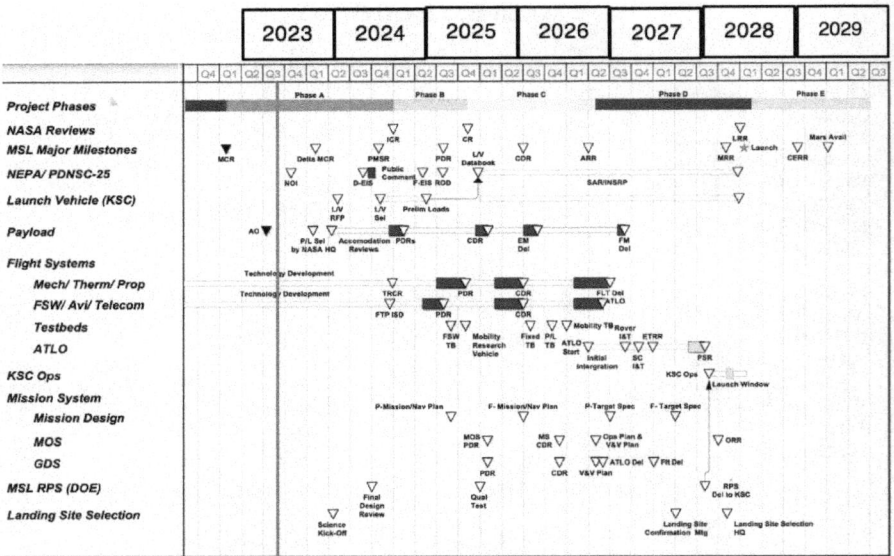

Fig. 23 Schedule

Table 8   2028/2029 Mars Launch Windows

LAUNCH	DURATION	ARRIVAL	DELTA-V (km/s)
12/31/2028	240	8/28/2029	5.17
12/31/2028	224	8/12/2029	5.2
1/16/2029	192	7/27/2029	6.06
1/16/2029	176	7/11/2029	6.73

REFERENCES

1.Gilbert V. Levin and Patricia Ann Straat· The Case for Extant Life on Mars and Its Possible Detection by the Viking Labeled Release Experiment, Astrobiology. 2016 Oct 1; 16(10): 798–810

2. A Chiral Labeled Release Instrument For In Situ Detection Of Extant Life. A. D. Anbar[1] and G. V. Levin[2] , 1 School of Earth & Space Exploration, Arizona State University, Tempe, AZ , 2 Beyond Center, School of Liberal Arts and Sciences, Arizona State University, Tempe, AZ

3. Kaplan, M. DNA has a 521-year half-life. *Nature* (2012). https://doi.org/10.1038/nature.2012.11555

4."Mining" Water Ice on Mars An Assessment of ISRU Options in Support of Future Human Missions Stephen Hoffman, Alida Andrews, Kevin Watts July 2016

5. Human Exploration of Mars Design Reference Architecture 5.0, NASA/SP–2009–566

6. Kaczur JJ, Yang H, Liu Z, Sajjad SD and Masel RI (2018) Carbon Dioxide and Water Electrolysis Using New Alkaline Stable Anion Membranes. Front. Chem. 6:263. doi: 10.3389/fchem.2018.00263

7. David I. Poston, Marc A. Gibson, Rene G. Sanchez & Patrick R. McClure (2020) Results of the KRUSTY Nuclear System Test, Nuclear Technology, 206:sup1, S89-S117, DOI: 10.1080/00295450.2020.1730673

8. Mikhaylik, Y., 650 WH/KG Rechargable Batteries, *2018 NASA Aerospace Battery Workshop*

9. Greenbaum, C. Mars Experimental Aquaponics Laboratory System (MEALS), 2021 Mars Society Convention

10. Przybyla, C. Space Aquaculture: Prospects for Raising Aquatic Vertebrates in a Bioregenerative Life-Support System on a Lunar Base

11. Harrison, D.A. et al, Next Generation Rover for Lunar Exploration, NASA-JSC

12. J. Balaram, Theodore Tzanetos, Håvard Fjær Grip Jet Propulsion Laboratory, California Institute of Technology, Mars Science Helicopter Conceptual Design, NASA/TM—2020–220485

4: TELEROBOTIC MARS MISSION FOR LAVA TUBE EXPLORATION AND EXAMINATION OF LIFE

Hanjo Schnellbaecher
Hanjo.Schnellbaecher@TUDSaT.space
Florian Dufresne
Dufresne.Florian@gmail.com
Tommy Nilsson
Tommy.Nilsson@ESA.int
Leonie Becker
Leonie.Bensch@DLR.de
Oliver Bensch
Oliver.Bensch@DLR.de
Enrico Guerra
EnricoGuerra@outlook.com
Wafa Sadri
Wafa.Sadri@ESA.int
Vanessa Neumann
Vanessa.Neumann@TUDSaT.space

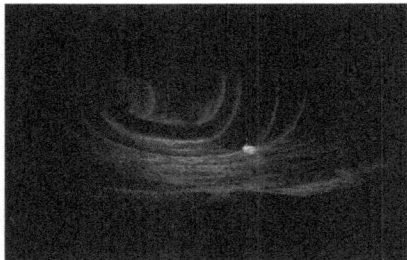

AIM AND GENERAL PHILOSOPHY

The general profile and overarching goal of our proposed mission is to pioneer potentially highly beneficial, or even vital, and cost-effective techniques for the future human colonization of Mars. Adopting radically new and disruptive solutions untested in the Martian context, our approach is one of high risk and high reward. The real possibility of such a solution failing has prompted us to base our mission architecture around a rover carrying a set of 6 distinct experimental payloads, each capable of operating independently on the others, thus substantially increasing the chances of the mission yielding some valuable findings. At the same time, we sought to exploit available synergies by assembling a combination of payloads that would together form a coherent experimental ecosystem, with each payload providing potential value to the others. Apart from providing such a testbed for evaluation of novel technological solutions, another aim of our proposed mission is to help generate scientific know-how enhancing our understanding of the Red Planet.

Mars has been attracting scientific attention predominantly as the most likely planet to provide direct indication of life beyond Earth [1] as well as for its potential habitability [2]. While several robotic missions seeking to find signs of Martian life have already taken place (e.g., Curiosity), substantial areas of the Martian landscape remain unexplored. Chiefly, research indicates that lava tubes on Mars might provide conditions particularly conducive to life, due to stable temperatures and shielding from radiation [3]. Of equal interest is the exploration of conditions that might support life on Mars in the future. Developing reliable strategies for plant growth, for instance, will likely prove crucial for future Martian outposts. By way of example, studies on Earth have shown that certain species of fungi can thrive in extreme environments and even develop resilience to high levels of radiation [4]. Our ability to understand and take advantage of such opportunities might prove indispensable for humanity's future colonization of Mars.

To this end, our mission takes aim at the Nili-Fossae region, rich in natural resources (and carbonates in particular), past water repositories and signs of volcanic activity. With our proposed experimental payloads, we intend to explore existing lava-tubes, search for signs of past life and assess their potentially valuable geological features for future base building. We will evaluate biomatter in the form of plants and fungi as possible food and base-building materials respectively. Finally, we seek to explore a variety of novel power generation techniques using the Martian atmosphere and gravity. As detailed throughout the remainder of this report, this assemblage of experimental payloads, then, constitutes the backbone of our proposed telerobotic mission to Mars.

LOCATION

Our search for a preferred landing site was primarily guided by two goals: maximizing the potential for scientific yield and minimizing environmental risks posed to the mission.

In terms of scientific yield, the regions on Mars deemed to be of greatest interest are the ones believed to once have contained water, hence being more likely to have preserved signs of past life [5]. A range of such sites scattered across the planet has been identified through previous research [6].

In terms of risk mitigation, the Martian southern hemisphere is mountainous, covered by complex terrain, making it difficult for rovers (or any other surface technology) to safely move around. In contrast, the northern hemisphere is considerably flatter, consisting mostly of lowlands. This makes navigation both easier and safer. Prior rover missions were targeting the northern hemisphere for this reason.

Drawing on such past missions and existing research, we determined that the Nili Fossae region would be the most suitable landing site. Located in the northern hemisphere, not only is this region comparatively safe and accessible for exploratory activities, but it is likewise believed to have once been rich in water [7] and consequently represents a suitable location for the search of life. Furthermore, as elaborated in the following subsection, evidence of past volcanic activity suggests that Nili Fossae might likewise be home to unexplored lava tubes [8].

Lava Tube Exploration

Future human exploration missions to Mars will inevitably entail significant challenges, ranging from high radiation levels, micrometeoroids to general harsh climate conditions on the Martian surface [9]. One possible solution to mitigate these adverse environmental conditions on the red planet is the use of lava tubes that have formed naturally beneath the surface as shelter for human exploration. In this regard, [9] concluded that "Caves may be among the only structures on Mars that offer long-term protection from such hazards." (p. 1).

The structures do not only protect potential astronauts from radiations, but the ambient temperature inside lava tube structures remains constant, which would yield important advantages in comparison to highly fluctuating and extreme temperatures on the Martian surface [3].

Scenario	Exposure Time	Cumulative Dose (mSV)
Surface Mission	24hr/day	14.795
Cave Habitat	24hr/day	0.012

Table 1: Radiation Dose Comparison Mars Surface vs. Cave Habitat. Reprinted from [8].

Lava tubes were formed by the flow of low viscosity basaltic lava during the eruption of a non-explosive volcano [10]. Due to the lower gravity conditions, the diameter of lava tubes on Mars is thought to be several hundreds of meters wide [8], with walls tens of meters thick [11].

Yet, the exploration of lava tubes would not only be important with regards to future human settlements, but also from a research perspective, as life forms would also be protected inside the caves from hazardous conditions on the Martian surface [3]. Importantly, water resources could have been trapped and protected by the environment inside the tubes [11]. As lava tubes are generally located under the Martian surface, they would also yield a unique opportunity to scientifically explore minerals and material samples that are not accessible from the surface. The analysis of material probes that have been protected inside lava tubes could therefore also yield insights into the developmental history of Mars in terms of volcanic and thermal activity [3]. Similar missions have also already been investigated: For instance, the NASA-funded "BRAILLE" (Biologic and Resource Analog Investigations in Low Light Environments) project explores potential use of analog cave environments as a training ground for future lava tube missions. Simulation of scenarios involving search for lifeforms as well as life-sustaining minerals is of particular interest.

Based on these findings, we propose that one of our mission goals is the exploration of lava tubes to locate possible water resources, minerals and possibly even lifeforms that have been protected inside the structures, as this is a highly relevant research area with a considerable potential for scientific yield and value.

Lava Tube Locations
Several different locations for lava tube entrances have been identified and mapped. In this regard, for instance, seven skylight entrances could be observed around the Arsia Mons using the orbiting Thermal Emission Imaging System (THEMIS) of the Mars Odyssey orbiter, which uses infrared imaging to detect subsurface structures, or the HiRISE camera on the Mars Reconnaissance Orbiter, which operates in visible wavelengths [12], [9]. As for the Nili Fossae region - the proposed landing site for our mission - it is located in the Volcanic region of Syrtis major and consequently considered likely to be the home of lava tubes under its surface [8]. Actual skylight entrances of lava tubes around Nili Fossae have yet to be spotted using data from the Mars Odyssey Orbiter though. As our mission is oriented towards scouting and exploration, the risk of placing our mission in Nili Fossae without finding any lava-tubes may be acceptable, especially considering that most of the technologies and experiments would still provide critical data (e.g. Greenhouse (see sec. 8), wind balloon (see sec. 5) ...). A location like Arisa Mons, for which lava tube entrances have already been spotted, may also serve as a backup location in case the Mars Reconnaissance orbiter would fail at finding lava tubes entrances near Nili Fossae within the 10-year window that remains till mission launch.

The planned payloads for the telerobotic rover mission are proposed and explained in detail in the following sections. In this regard, each subsystem's description will include an overview of the engineering design, as well as the value for exploration preparation and scientific return. In addition, estimated costs for each payload and a schedule will be provided.

Cost estimations for development and integration are based on the average NASA salary of 124.363$ [13]. As defined by NASA's Procedural Requirements NPR 7120.5F [14], the payloads were divided into subsystems according to NASA's Work Breakdown Structure (level 3) [15] and costs were calculated for each subsystem individually. Total costs for all payloads (WBS level 2) for the Project Life-Cycle Cost Estimate (LCCE) are shown in section 11.

The rover platform design will be presented after the presentation of six different payload subsystems, followed by a description of an overall mission schedule and timeline based on a project life-cycle model.

PAYLOAD 1: GROUND PENETRATING RADAR & CAMERA SYSTEM

Engineering Design
The proposed rover will use the Mastcam-Z, that is also attached to the Perseverance rover that was launched during the Mars 2020 mission. As was the case in the Mars 2020 mission, the camera will be attached to the body of the main rover [16]. For a detailed description of the system characteristics associated with the Mastcam-Z system, please refer to [16]. We plan to reuse the Mastcam-Z camera, despite its 2-megapixel camera, since it was constructed with the intent to deliver as much information as possible, with limited bandwidth available.

The ground penetrating radar system will be based on the RIMFAX system that was proposed by NASA [18] and utilized for the Perseverance rover mission. The technology can scan up to a depth of around 10m. Please refer to table 2 for a detailed description of all the system characteristics.

Location	The radar antenna is on the lower rear of the rover
Mass	Less than 6.6 pounds (3 kilograms)
Power	5 to 10 watts
Volume	7 by 4.7 by 2.4 inches (196 x 120 x 66 millimeters)
Data Return	5 to 10 kilobytes per sounding location
Frequency Range	150 to 1200 megahertz
Vertical Resolution	As small as about 3 to 12 inches thick (15 to 30 centimeters) thick
Penetration Depth	Greater than 30 feet (10 meters) deep depending on materials
Measurement Interval	About every 4 inches (10 centimeters) along the rover track

Table 2: RIMFAX Ground Penetrating Radar System Characteristics. Reprinted from [18].

Location	Mounted on the rover mast at the eye level of a 6 ½-foot-tall person (2 meters tall). The cameras are separated by 9.5 inches (24.2 centimeters) to provide stereo vision.
Mass	Approximately 8.8 pounds (about 4 kilograms)
Power	Approximately 17.4 watts
Volume	Camera head, per unit: 4.3 by 4.7 by 10.2 inches (11 by 12 by 26 centimeters) Digital electronics assembly: 8.6 by 4.7 by 1.9 inches (22 by 12 by 5 centimeters) Calibration target: 3.9 by 3.9 by 2.7 inches (10 by 10 by 7 centimeters)
Data Return	Approximately 148 megabits per sol, average
Color Quality	Similar to that of a consumer digital camera (2-megapixel)
Image Size	1600 by 1200 pixels maximum
Image Resolution	Able to resolve between about 150 microns per pixel (0.15 millimeter or 0.0059 inch) to 7.4 millimeters (0.3 inches) per pixel depending on distance

Table 3: Mastcam-Z Camera System Characteristics. Reprinted from [17].

Exploration Preparation

Lava tubes have so far only been explored from above the surface by mapping the cave entrances using data from Martian orbiters, as briefly mentioned previously. As a result, mapping and exploring a lava tube with a rover mission would be the next step in gaining a better understanding of these important structures. Drawing on the orbiter's location data, characteristics, such as the shape of a lava tube's entrance, could subsequently form the basis for decisions concerning further exploration using micro-rovers stored as a payload in the main Martian rover.

The rover will be able to drive to the entrance from a nearby landing point. The Mastcam-Z camera system attached to the main rover can send images of the tube entrances to Earth, where experts can assess the cave's exploration value. Large caves with an entrance and structure that allow for human exploration and habitation, for example, would constitute an ideal mission location.

Furthermore, the RIMFAX ground penetrating radar system could estimate the size and depth (up to 10 - 20m, [19], [18]) of the tube from above the surface, assisting experts in their decision-making process.

After identifying and selecting the lava tube's entrance through the inspection completed by the main rover, a detailed inspection of lava tubes could follow through our rover mission, including the deployment of small robot swarms inside the lava tube.

Scientific Return

Even though both systems have already been utilized for the Perseverance rover mission, the Mastcam-Z and RIMFAX can both yield important additional insights due to the distinct landing location of our rover and the unique environment inside the lava tubes.

Overall, the Mastcam-Z camera system yields important additional visual information about the Martian environment. Camera data can likewise be used to conduct a spectral analysis of various rock samples on the surface of Mars to aid mineral identification. Here, [20] and [16] suggested that camera data collected by the proposed camera system can successfully be used to estimate different minerals. Furthermore, camera data can, for instance, be used to monitor the terrain and atmosphere of the Martian environment (See table 4 for a detailed description of various camera objectives, please refer to [16].)

Mastcam-Z goals	Mastcam-Z detailed investigation objectives
1. Characterize the overall landscape geomorphology, processes, and the nature of the geologic record (mineralogy, texture, structure, stratigraphy) at the rover field site	1-a. Characterize the morphology, texture, and multispectral properties of rocks and outcrops to assess emplacement history, variability of composition, and physical properties.
	1-b. Determine the structure and orientation of stratigraphic boundaries, layers, and other key morphologic features to investigate emplacement and modification history.
	1-c. Characterize the position, size, morphology, texture, and multispectral properties of rocks and fines to constrain provenance and weathering history.
	1-d. Observe and monitor terrains disturbed by rover wheels and other hardware elements to assess surface to physical and chemical weathering.
	1-e. Distinguish among bedform types within the vicinity of the rover to evaluate the modification history of the landscape.
	1-f. Identify diagnostic sedimentary structures to determine emplacement history.
	1-g. Characterize finer scale color/spectral variation (e.g., cm-scale veins, post-depositional concretions) to constrain provenance and diagenetic history.
2. Assess current atmospheric and astronomical conditions, events, and surface-atmosphere interactions and processes	2-a. Observe the Sun for rover navigation and atmospheric science purposes.
	2-b. Observe the sky and surface/atmosphere boundary layer to measure atmospheric aerosol/cloud properties and transient atmospheric/astronomical events.
3. Provide operational support and scientific context for rover navigation, contact science, sample selection, extraction, caching, and other Mars 2020 investigations	3-a. Acquire stereo images for navigation, instrument deployment, and other operational purposes on a tactical timescale.
	3-b. Acquire sub-mm/pixel scale images of targets close to the rover.
	3-c. Resolve morphology and color/multispectral properties of distant geologic features and topography for longer-term science and localization/navigation planning purposes.

Table 2: Mastcam- Z Objectives. Reprinted from [16].

The ground-penetrating radar system could additionally be used to identify water resources on Mars. [21] and [22] proposed that using ground-penetrating radar to map such structures on Mars could be as successful as it is on Earth. The radar system in combination with camera data could also be used later in the mission to investigate potential water-ice sources [19] and could yield important insights into the ancient development of the Martian surface [18].

Costs
As the development of the Mastcam-Z and the RIMFAX system are already completed, only adjustments to the hardware have to be made. No information about the specific costs of the Mastcam-Z system and the RIMFAX system appears to be available. However, information about sub-components can be found or inferred. E.g., the optical sensor used in the Mastcam-Z KAI-2020 by Kodak was listed by vendors for about $3500. Prices for the other components cannot be found directly and can therefore only be broadly estimated. The RIMFAX system is separated into two modules, the processor BAE RAD6000 which costs about $300.000, and the ground low-gain x-band antennae manufactured by enduroSAT. No prices were released by enduroSAT, nevertheless prices for ground penetrating sensors can go up to $50.000. Please refer to table 5 for a detailed estimation of costs for the camera system, and to table 6 for the RIMFAX system based on information found online.

Component	KAI-2020 optical sensor (Kodak)	Lenses	Other components	Development & Integration	Total
Cost estimate ($)	$3500	$50.000	$40.000	$100.000	$193.500

Table 3: Cost Estimate ($): Mastcam-Z.

Component	BAE RAD6000	Ground Sensor	Other components	Development & Integration	Total
Cost estimate ($)	$300.000	$50.000	$50.000	$100.000	$500.000

Table 4: Cost Estimate ($): RIMFAX.

Schedule
As the Mastcam- Z and the RIMFAX ground penetrating radar system have already been planned to be used in the Mars 2020 missions, it is ensured that the goals of the missions as well as the development process of the technologies will be completed before 2033.

PAYLOAD 2: LAVA TUBE EXPLORATION ROBOTS

Engineering Design
Prior research into potential Mars exploratory vehicles has presented various robot designs, ranging from flying drones [23] to autonomous mini-rovers [24] to hopping or rolling robots [25], [26]. However, in the context of Martian lava tube exploration, the German Aerospace Center (DLR) proposed an extremely robust robot designed specifically for the challenging environment (such as debris inside the tube) [27]. The snake-like robot is made up of various modules, each with rimless wheels attached to the side. Each robot is made up of a single main module that houses a power and communication source.

The side modules have their own actuators and can carry up to 6kg of payload. A flexible connector between the various modules provides shock absorption and flexibility, allowing the robot to be dropped inside the tube from a height of approximately 1.5m. If one of the rimless wheels becomes stuck inside the debris, it can be automatically removed from the robot's body and the exploration mission can continue as planned [27].

To collect data from inside the tube, the robots can be equipped with cameras and spectroscopic technology. Furthermore, various mapping technologies could be used to map the inside of the tube, such as LIDAR systems. However, using the space inside the rover's side elements, it could be outfitted with other tools for analyzing minerals or collecting samples. The samples could then be transported back to the cave's entrance and analyzed inside a station lowered inside the cave and containing the gas chromatograph instrument (see section 7), which could also serve as a charging and communication station for the rovers (see gravitational battery station system). Multiple robots could be lowered inside the tube to map the environment efficiently and without overlap while communicating using a SLAM-based algorithm ([28].

Figure 1: DLR Lava Tube Scout Robot based on [27].

Rover mass	~18kg
Payload capacity per aux module	6kg, 5l volume
Max. speed	~1.7 m/s
Max. obstacle height	>400mm
Tested drop height	>1.5m
Battery run time (single/ double)	>5h/10h
Number of modules	2-5

Table 5: Scout Rover Characteristics. Reprinted from [27].

Exploration Preparation

The topology of the lava tubes raises as a key interrogation if they are considered as potential anti-radiation shelters for astronauts living on Mars. Mapping their interior through LIDAR sensors or understanding their geological nature is a critical prerequisite for initiating any habitat design activities or focusing on other solutions with respect to astronaut protection. Such data could even be used to recreate the lava tubes in immersive virtual reality to facilitate training of future astronauts going to Mars.

Scientific Return

The scouting rover shall be able to grab geological samples deep in the lava-tube and bring them back to an analyzer. Such samples would, in turn, facilitate greater understanding of the geological nature and history of this volcanic region. Additionally, this procedure might enable us to analyze possible water samples and minerals located inside the lava tube. As each element of the robot can contain up to 6kg of payload, each robot can collect several separate samples inside the structure without contamination risks between samples. Therefore, the robot design allows for maximum science return while avoiding unnecessary risks by minimizing travel distance.

Costs

As the costs for the development and the manufacturing of the DLR scout rover has not been made available to the public, no definite price for the system can be determined. However, as the development of the robotic platform is already in an advanced stage, we can assume that the costs for further development will be comparatively low.

Costs for NASA's Mars rovers, on the other hand, have been made public. Based on the DLR scout rover's complexity and size, we can deduce that the development costs of NASA's Pathfinder rover of $174,2 million would be roughly comparable.

Schedule

Importantly, the proposed rim-less scout rover concept is already in advanced stage of development, with the concept scheduled to be mission ready in 2030. The ARCHES analog mission, a collaboration between DLR and ESA, will conduct initial tests this year inside a lava tube on Earth [27].

PAYLOAD 3: WIND POWER BALLOON

Figure 2: Wind Power Balloon Illustration.

The underlying concept here is very simple: A wind turbine similar in shape to a turbine of a passenger jet-plane but with a vastly bigger size, where the tube-shaped outside is an inflatable balloon. Alongside the turbine, multiple scientific payloads, provided the buoyancy permits it, can also be mounted onto it.

Engineering Design

The balloon will be anchored to the ground by an extensible power cord, allowing for the adjustment of altitude of the balloon. Mounted inside the tube sits the turbine, with rotors and the generator, with a significantly lower weight compared to a similar terrestrial turbine, due to the atmospheric density being substantially lower. Next to the turbine it is also conceivable to attach various scientific instruments to monitor atmospheric composition, pressure, temperature, and particles, such that are being flung around during Martian storms. Finally, such a platform could wear a coating made of electrodes to allow electrons from the ground to flow through the material of the tether, due to the difference of electric potential between the ground and the relatively high potential of the Martian atmosphere, thus creating an electric current.

The characteristics of the Martian atmosphere provide unique challenges and opportunities compared to flying a balloon on Earth. The biggest challenge is by far the extremely low density, making it more challenging to achieve buoyancy, when carrying heavy equipment, such as the turbine or a scientific payload, requiring the lifting volume to be scaled up. On the flip side, the low density also equals a lower force of wind, where a storm on Mars is comparable to a summer's breeze, lowering the danger of damage even in storms and allowing for lighter materials in the anchoring, hull of the balloon and turbine but also lowering the expected power output. The temperature is also a notable difference, ranging from about 20C° to around - 100C° at its coldest around equatorial regions. This should be kept in mind when selecting equipment, material and lifting gas.

Speaking of lifting gasses: One major advantage of the Martian atmosphere over the Terran one is that it consists almost exclusively out of CO_2, having a comparatively high molar mass of around 44g/mol. Any gas with a lower molar mass is a potential lifting gas, including the popular Helium (4g/mol), but also Oxygen (O_2 has around 32g/mol) and Hydrogen. A problem with Oxygen, however, is that it can be very reactive and corrode the hull over time, particularly when being transformed into Ozone by the very present radiation. Perhaps the biggest advantage of Oxygen is the in-situ availability, where it can be generated by either the greenhouse payload (see payload 6) or the already demonstrated MOXIE payload that could also be mounted on a big version of the balloon to autonomously regenerate its lifting gas anywhere. With an anchoring mechanism and a control circuit using the turbine for locomotion, such a balloon may serve as a permanently airborne independent drone, as proposed by [29].

Hydrogen is also a good candidate, which can be found on Mars in small quantities and is not flammable in the Martian atmosphere but has the downside of easily escaping the lifting volume, due its elusiveness. Helium, whereas it has very favorable characteristics in reactivity, elusiveness and lifting capabilities, it will have to be brought entirely from Earth, as the Martian atmosphere is devoid of it in low altitudes. The balloon covering may be selected according to the lifting gas, but a black coloration for the absorption of radiation may be favorable, also in increasing the lift during the day due to the heating of the lifting gas by the increased absorption of sunlight. Table 8 lists a possible configuration of such a setup, with the assumption of a perfect cylindrical shape for the lifting volume, which ideally should be aerodynamic to facilitate better wind-flow and using O_2 as the lifting gas. It should be noted that the final values are expected to vary based on selected payloads and lifting gas.

The overall density in this example is lower than the Martian atmospheric density at about $0,02kg/m^3$, resulting in a positive buoyancy.

A detachable design has also been considered with an anchor, that can grapple the base station or attach itself to Martian rocks or even into the soil. Since the balloon has its own power source and a built-in turbine, it is conceivable for it to be an independent drone with a flight range and duration tied to its capability to retain or regain its own lifting gas. This however would make the drone more complex especially if it is supposed to have controls and avionics. The feasibility at this stage will have to be investigated further and may prove too challenging due to its engineering requirements.

Variables	Value
Outer radius (m)	7
Inner radius (m)	3
Length of tube (m)	6
Lifting gas density (kg/m^3)	0.008008584
Surface area weight (kg/m^2)	0.01
Tether length (m)	40
Tether weight per length (kg/m)	0.01
Scientific payload weight (kg)	2
Windmill weight (kg)	2

Table 6: Wind Power Balloon Configuration Set-Up.

Exploration Preparation
The turbine and atmospheric electricity-based generator will provide another option of power generation for Martian infrastructure as well as an option to monitor the environment from an intermediate altitude, allowing for payloads requiring a higher proximity for their measurements. It also provides a test environment for buoyancy-based aviation that can be controlled and tested over longer periods of time from the base itself. Both the power generation by wind and transportation aspects have been around for a long time on Earth but have never been tested on Mars. This mission not only makes both possible at once, but also makes wind-power more feasible by requiring minimal supporting infrastructure for variable altitudes and a way to funnel the very low intensity Martian winds into the turbine, while exploiting an alternative power source in the form of atmospheric electricity simultaneously.

Scientific Return
The primary scientific return of the balloon will stem from its monitoring and observations of the atmosphere. While it may rest at high altitudes by default, ultimately the balloon can be put at desired altitudes for different tests. Results, especially during sandstorms, might be of particular interest. Moreover, in terrestrial environments, it is feasible for a small corona motor to be powered by exploiting the electric potential of the atmosphere. It could then be rewarding to verify if the same principle would work on Mars under fair-weather conditions. Additionally, the balloon may be employed to assist with scouting and locating suitable lava tube entrances, beginning shortly after deployment on Mars.

Costs
The cost is difficult to gauge due to variabilities in the possible setup. In the following table an estimate of costs for the most complex configuration is proposed. Since this subsystem is still in its early stages of development, it can be assumed that salaries will be the biggest expense during the development and integration phases over several years. It should be noted here that most of the cost depends on the scientific payloads used.

Element	Lifting gas (O2 tank)	Surface covering	Tether system	Turbine system	Scientific payloads	Development	Integration	Total
Cost estimate ($)	350	1000	500	500	5000	2.000.000	100.000	2.107.350

Table 7: Estimated Costs - Wind Power Balloon System.

Schedule

While it is not easy to pinpoint a schedule based on previous missions or similar experiments, the principles and engineering are very simple and most parts, except for the custom-made hull and an adjusted inflation-system, can be bought off the shelf with minimal adjustments. Since it also does not need to be pressurized during transport, the storage-equipment can also be very minimal. Therefore, it can be argued that a launch by 2033 is very reasonable as the required development should be rather minimal.

PAYLOAD 4: MYCOTECTURE

The idea of mycotecture (a combination of the words *Mycelium*, the "root" part of fungi, and *architecture*) represents a relatively novel approach to producing building material and for building, or rather growing, structures. Here the properties of fungi to "mold" into any shape, filling it completely with fungal mass that is sturdy but also relatively light weight while only requiring oxygen, water, and nutrients, is being used to create partial or entire structures. This concept has been investigated by both NASA [30] and ESA [31] among others [32].

Engineering Design

To provide the growth environment, the following resources need to be available: A mold or inflatable structure for the mushroom to inhabit, water, nutrients, oxygen, and the right thermal conditions.

The mold can come in two different forms: A rigid block to create one or multiple uniform bricks of mycelium or an inflatable structure that can then be filled out by the mycelium. A hybridization of both is also thinkable.

Water, nutrients, and oxygen can come pre-processed from earth as part of the setup or be possibly sourced from the greenhouse payload (see section 7). Whereas the latter option forms a nice synergy with the greenhouse payload, the former option should always have preference for the sake of redundancy. Oxygen may successfully be sourced from the greenhouse payload (see payload 6), should its deployment reach a significant level of success.

Thermal control might constitute the biggest challenge, one that is not widely discussed in existing literature, which seems to generally presuppose optimal temperature. One obvious requirement is for the growing environment not to freeze over. How low the temperatures can go without damaging the fungi is however difficult to assess. On the other hand, low temperature can be exploited for "deactivating" the mycelium simply by exposing it to the Martian climate without providing additional heating.

A potential mycotecture system could have these estimated characteristics during transport:

Mass	Volume	Energy consumption
50kg	<1m^3	~3W, if insulated well

Table 8: Mycotecture Payload Engineering Design.

Exploration Preparation
The notion of growing fungi in off-world conditions has seen a surging interest in recent years, with relevant studies ongoing. Our overall aspiration with this technology is for it to provide shelter and accommodations for future Martian inhabitants. The risk of these attempts failing, however, are considerable. Therefore, it is crucial to assess mycotecture solutions thoroughly, proving that they are viable on Mars before they are relied on during subsequent crewed missions. In addition, fungi are known to be able to extract minerals from soil, which will also be crucial for foraging vital resources for many purposes of Mars colonization. In the light of the great potential and the great unknowns still surrounding this technology, we can safely say that there is a strong incentive to studying it further.

Scientific Return
The main scientific interest with fungi in the context of this mission will be to examine techniques under which fungi, specifically mycelium, can grow. This includes figuring out ideal conditions on Mars for growth as well as trying to use in-situ resources, such as soil, oxygen from photosynthesized or electrolyzed carbon-dioxide and possibly water from methane and oxygen, if available. This all links to the very central question of how life can thrive on Mars.

Costs
Since there is no suggestion on a concrete design for a mycelium growing system for Mars, this estimate tries to give a rough idea with generous margins included in the calculation. To try as many different techniques as possible, both an inflatable system and a solid one will be covered. The mycotecture will use a similar setup as can be found in the greenhouse payload. This is to reduce development cost, as it can be reasonably assumed that the mycelium thrives in a similar, if not the same environment, as the greenhouse plant. Due to the early development stage of this subsystem, the biggest costs are expected in HR for development and integration over several years.

Element	Grow volume (solid and inflatable)	Resource supplementation syst.	Nutrients	Temp. control syst.	Redundancy	Development	Integration	Total
Cost estimate ($)	1500	1500	500	100	500	2,000,000	100,000	2,104,100

Table 9: Costs Estimates for the Mycotecture Payload.

Schedule
The technology behind mycotecture may be young but has been, and currently is, thoroughly investigated, having even been trialed on the ISS. As of today, no trial has however taken place on Mars. This means that there still is some development work required, but a launch by 2033 is still realistic, considering the current efforts.

PAYLOAD 5: GAS CHROMATOGRAPH

Engineering Design
As previously mentioned, lava tubes are more likely to host life or at least hold some traces of past life on the Red Planet. The Curiosity rover was looking for traces of life using the SAM (Sample Analysis at Mars) tool and our team plans to add the same kind of instrument to make sure our rover can track life forms in Lava tubes. Therefore, a gas chromatograph will be added to analyze the atmosphere inside the lava tubes. This technology does not present a substantial engineering challenge as it has already been designed and sent to Mars with Curiosity [33]. The characteristics of such an instrument may be as follows: A mass of 35kg, dimensions of 46cm x 27cm x 29 cm and a required power of 120W.

Exploration Preparation
Analyzing the inner atmosphere of lava tubes may be critical to the explorations task of the astronauts. Indeed, this may help to spot dangerous gasses that would appear in high concentration, in particular explosive ones.

Scientific Return
One of the gasses that is released by lifeforms is methane and for now only small concentrations of it have been detected in the Martian atmosphere by the rover Curiosity [34]. Potentially significant concentrations of methane have also been detected over the Nili Fossae region by Earth-based telescopes [35][36]. The origin of this methane may also be geological [37], and our mission may answer the question of its origin by bringing the Gas Chromatograph to Mars again, and more specifically to Nili Fossae.

Costs
Concerning the price of such an instrument, we will rely on the technology developed for the Curiosity rover. The provided costs estimate would then only include the acquisition of the components. Based on current market prices, we estimate this would translate into approximately 50,000$.

Schedule
Finally, this technology is already available and ready for deployment. Only the integration time and adaptation of the instrument to get samples from the lava tube exploration robots (see section 4) may require more time. However, those tasks would only require a few years, mostly for the integration and to create the anchor point for the robots. Such a development is reasonably achievable for a launch by 2033.

PAYLOAD 6: DEPLOYABLE GREENHOUSE

Figure 3: Concept for the deployable greenhouse

Engineering Design

The next payload that we are planning to bring to Mars is a Greenhouse that would be deployed on the surface to test full soil growth of plants. This experiment has a notable synergy with the mycotecture payload, as it will be able to supply the mycelium (payload 4) with oxygen which is critical for their growth. Two main elements are required for this payload: first, an automated greenhouse that can deploy and plant seeds in the Martial soil, and subsequently support the growing plants. The second element is a set of seeds from species whose growth would already have been tested into Mars soil simulants to give them the best chances to survive and grow.

Regarding the hardware setup, our design will mostly rely on what the open-source community came up with in the context of the Farmbot initiative [38]. It consists of a robotic autonomous farm that has a growth area as big as $18m^2$. Through its raspberry computer, it can monitor the growth of the plants while providing them with everything they need. The possibilities are even wider as some projects use the Farmbot in association with photogrammetry technologies to have a 1-to-1 digital twin of the garden. What we intend to add to this setup is a life support system (temperature control, atmosphere monitoring ...) to allow the greenhouse to survive Martian nights, but also some redundancy in case of system failures.

An area of 2x1m is conceivable to give enough space to the plants. The structure would then be 1.5m tall. The payload fits in a volume of 1x2x1.5m or 3 cubic meters. Once deployed on soft ground, the bottom part of the structure could start collecting the Martian soil to put it inside the greenhouse thanks to a bucket-wheel excavator. This technology appears to be a reliable choice for such a mission, as presented by the trade-off study realized by García de Quevedo Suero et al in the context of a Martian greenhouse-rover design project [39]. The farm bot would then spread the dirt on the whole surface, using a rake tool, from the movable platform that brought it in the grow volume (see figure 4).

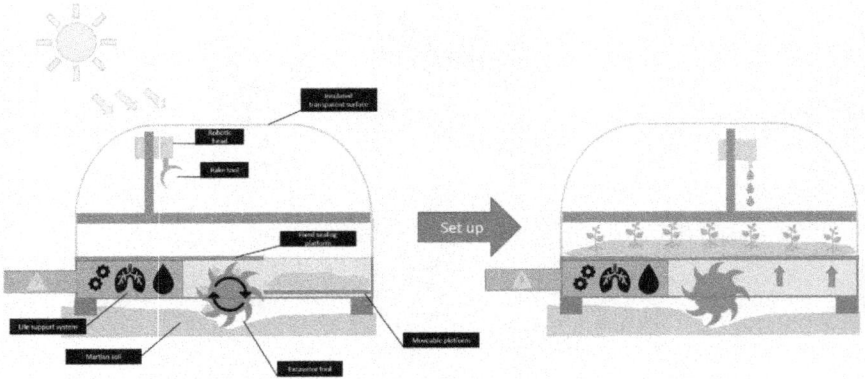

Figure 4: Greenhouse Design Concept, during (Left) and after (Right) Set-Up.

The Farmbot's energy consumption is estimated to be 0.287kWh/day by its manufacturer. In addition, to handle the temperatures that can drop to -73°C during the night, the life support system should be able to maintain the temperature around 20°C. The design results are reported in the table 12. Regarding the power required to maintain the temperature in the greenhouse, extensive efforts may have to be done on the dome insulation, using insulated glazing either by Mars atmosphere or by vacuum. These technologies would provide even lower heat transfer coefficients than the one that has been used for the design here (air insulated glazing).

Glassed Area	Double glazing heat transfer coeff.	Target temperature	Night temperature	Estimated heat transfer	Night Duration	Required energy /night
~5m²	~1.1 W/m².K	20°c	-73°c	511.5 W	12h20mins	6.3 kW.h

Table 10: Greenhouse Engineering Design.

The energy consumption of the excavator has not been addressed here because all other systems would be shut down during the excavation process and following the current design it will have enough power to run. Looking at the species to bring, studies have shown that nitrogen-fixer species, such as Lotus Pedunculatus, would enrich the soil with this critical nutrient for other species to grow. Moreover, some Crops species like Lepidium sativum are good candidates to grow on Martian soil. Both showed germination rates above 70% in Martian soil simulants [40]. Finally, the mass of the total setup may not go over 150kg, with approximately 30kg reserved for the Farmbot and 120kg for the structure and life support system.

Element	Farmbot	Deployable feature	Glass	Temp control syst.	Redundancy	Development	Integration	Total
Cost estimate ($)	3.500	2.000	500	100	500	2.000.000	100.000	2.106.600

Table 11: Cost Estimate for the Greenhouse.

Exploration Preparation
This experiment would assess the possibility to grow plants in Martian soil, which will become critical to foster human exploration and colonization of this planet. If it works, it could also provide the other payloads of the mission with oxygen if required.

Scientific Return
This will be the first time that organisms from Earth grow on Mars. We may furthermore identify those species that are the most suitable for growing on Mars by bringing seeds from different plants that have been demonstrated to grow in Martian regolith simulants. However, the growth in regolith may not go as smoothly as expected [41] and failure would also give valuable data contributing to greater chances of success during future attempts. This is a unique chance for a full soil plant growing attempt. On the other hand, this experiment introduces a high risk of Martian soil contamination in case of failure of the greenhouse. Thus, planetary protection protocols must be followed rigorously to avoid any contamination.

Costs
Concerning the cost of such a set-up, Farmbot-wise, the 4.5m² version of the robot would cost around 3500$. However, our design would make the structural elements of the set-up deployable (e.g., using a dedicated crane) which represents an additional cost estimated at around another two thousand dollars, even if we plan to have a smaller growth area. There's also a need for an environmental control system and some redundancy. Since the system is in an early stage, the biggest costs could be expected for development and integration. The cost breakdown is reported in table 13.

Schedule
About the technologies' readiness, the Farmbot products should be available no later than summer 2022, and all other elements constituting the design we propose are already available on the market at reasonable costs. Once the Farmbot is received, we estimate that 4 years would be required to make the system Mars ready by adding all the aforementioned features. This leaves even enough time to try out this greenhouse design on the Moon first before sending it to Mars.

PLATFORM/SYSTEM (ROVER(S)/BASE(S))

Winch System and Gravity-Based Power Generation
This section deals with the engineering design of the winch system. This tool will allow our rover to go down into the lava-tube to set up the exploration robot swarm's station and then to go back up again. This system would be able to take

advantage of the gravitational potential energy turned into kinetic energy during descent, thus generating power as it travels down the tube's walls.

In terms of technology readiness, winch systems are already mature and commonly used on Earth. Their main components are a direct current motor, an electronic board to control it, some gear systems for power transmission, a shaft and a cable/rope wound around it.

The "worst-case" scenario we considered while designing this system was the rover running into problems while climbing during the descent. The rover should then be able to bring the payload back to the central station tethered at the top of the lava tube. The application of Newton's 2nd law to the payload-winch system leads to the results presented in table 14 below:

Payload mass	Ascend/Descent speed	Vertical distance	Required winch motor power (+10% margin)	DC motor round trip efficiency as generator	Generated energy per descent
500 kg	40 cm/s	100 m	0.746 (0.820) kW	70%	36,24 Wh

Table 12: Winch System Requirements and Power Generation Capabilities.

According to Boretti et al [42], in 2013 the round-trip efficiency, meaning from wheels to battery and wheels again, of a regenerative braking like the one we would like to use was estimated at most, around 70% for electric cars. Note that dedicated dynamo can have efficiencies up to 80%. It must also be considered that this estimation is highly dependent on how people drive the cars. In our case, however, we can assume that the rover goes down in a way that maximizes energy generation. This additional energy generation system won't clearly be enough to provide other systems with power, nonetheless it covers some of the power requirements of the winch system. Moreover, winch systems fitting those requirements may be found on the market with the characteristics reported in table 15.

Model	Price	Mass	Electric consumption	Dimensions (LxWxH)
L-GT300-SY-12V	~400$	17kg	1700kW	391mm x 126mm x 128mm

Table 13: Example of an off the Shelf Winch System Fitting our Requirements [43].

By adding additional redundancy (another motor) and the possibility to transmit data and power through the winch cable, as part of the central station energy generation endeavor, it must be feasible to design a similar system in terms of dimensions and electric consumption at an additional cost estimated around 300$.

Avionics and Sensors
The avionics unit will contain all the electronics of the rover that handles data and resources transmission between the different payloads. It will process information from the sensing organs of the rover that will be added: the 360° camera, thermometers, pressure sensors, and microphones. This unit is also responsible for

communication with Earth through an antenna. Particular attention may then be paid to the integration and physical protection of this part, this is the reason why it is planned to have it at the center of the main station deployed at the entrance of the lava tube. Furthermore, this is a strategic location in the sense that it will then be directly connected to all payloads. Embedded systems and their protection may weigh around one metric ton according to current rover designs (Perseverance [44]). The avionics unit is not able to handle temperatures out of the range -40°c/+40°C on the contemporary generation of rovers [45]. On Mars, the temperature may drop to -90°C during nights which would cause the rover to freeze to death without the proper equipment. Indeed, the electronics will release some heat over time, thus making use of gold paint, good insulation and heaters will help maintain the temperature of the components in the acceptable range.

Chassis and Navigation system
Concerning the navigation system, the proposed design does not reinvent the wheel: It will be equipped with a chassis similar to the current generation of Martian rovers with 6 motorized wheels, also known as the rocker-bogie design, that will allow it to move on the chaotic terrain of the lava tubes especially. The chassis will be modular, allowing the payloads to be deployed individually onto the Martian surface. This element will concentrate most of the rover's mass. It must also be mentioned that the chassis will have to be big enough to host all payloads in it, which corresponds to approximately 270kg and a volume close to 5 cubic meters.

Power Generation
This part focuses on the elements that will generate and store electricity. Some of the systems that will be carried can generate power: The winch system and the wind balloon. However, these are not producing enough power to supply all other systems in the mission. The elements that would require constant energy supply to work are the rover's "brains" (avionics), the greenhouse, the mycotecture and the drone control station. Thus, they are the one critical to the power supply design. The other components only need energy at one point, but we still need to make sure the rover's power generation system would provide them with enough power at the right time.

A Radioisotope Thermoelectric Generator (RTG) can produce a few hundred watts, like 110W for the Perseverance rover, which would be sufficient to cover our mission's power requirements. On Perseverance this unit weighed 45kg for 64cm diameter and 66cm length. This represents a source of energy that is independent of current lighting conditions, unlike solar panels. One of the drawbacks of that technology is that it releases heat, but this can be turned into an advantage if that wasted heat, which is not turned into electricity, could be used for thermal regulation purposes, particularly during nights.

To conclude on the platform design, it is hard to give an estimate on how much the chassis will weigh, but at this stage of our mission mass estimate, more than 9 tons are available to the platform design with respect to the lander capabilities, which should comfortably cover the needs of the chassis.

SCHEDULE: PROJECT LIFE-CYCLE

This project is now in phase "Pre-A" as per the NASA Procedural Requirements NPR 7120.5F [14] "Subject: NASA Space Flight Program and Project Management Requirements". The NASA project life cycle is comparable to the waterfall methodology [46]. Key Decision Points A (KDP A) must be established and approved by the project's decision authority to move on to phase "A" of the project. No phase will be repeated, in contrast to agile approaches like SCRUM [47].

This project's experiment payloads are the main topic of this work. Given the estimated schedule we provided for each payload, it has been determined that launching the mission by 2033 should be feasible. Based on historical data (e.g., schedule of Perseverance rover), it can be assumed that other processes, such as implementing flight safety standards in accordance with NPR 7120.5F, shouldn't cause schedule delays. A GANTT chart in table 16 shows an example timeline for all planned phases A through F with a launch at the end of phase D in 2033.

Phase	Life-Cycle Phases \ Year (20XX)	'22	'23	'24	'25	'26	'27	'28	'29	'30	'31	'32	'33	'34	'35	'36
Pre A	Concept Studies	██														
A	Concept Technology Development		██													
B	Preliminary Design			██												
C	Final Design Fabrication				██	██										
D	Assembly, Integration & Verification									██	██					
E	Operation Sustainment											██	██			
F	Closeout														██	██

Table 14: Project Lifecycle: GANTT Chart.

The primary mission (phase E) will have the following structure depending on the intended payloads:

1. Initial phase: The rover system will land on Mars on a relatively safe spot, which is in reasonably proximity of probable locations of lava tube entrances. After landing, routine self-assessment and tests will be commenced in conjunction with the deployment of mobile means of power generation and reconnaissance. This includes solar panels, the wind-turbine balloon and lava tube exploration robot swarms.

2. Transit phase: Next the rover will have to relocate to a lava-tube. It will do so by driving there, surveying the environment. Navigation is intended to be done by GNSS, if available at that time, or similar means, with a preselected location based on already available observations of Martian geography.

3. Settlement Phase: This phase encompasses two major components: The setup and usage of the greenhouse and mycotecture payloads as well as the lava tube exploration. This is done by setting up the base station at the entrance of the lava tube and by deploying the greenhouse along with the mycotecture environment. From there the winched rover will travel inside the lava-tube carrying the exploration robot swarm. These will subsequently map and survey the lava-tube.

By remobilization of the greenhouse and mycotecture payload, the Transit and Settlement phase can be looped to explore multiple lava-tubes in the area.

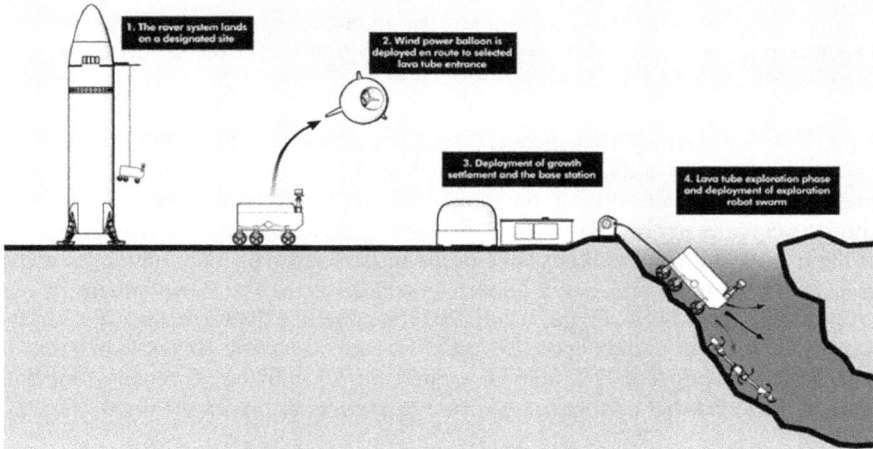

Figure 5: Schematic Mission Plan Illustration.

PROJECT LIFE-CYCLE COST ESTIMATE (LCCE)

We have provided cost breakdowns for several components. However, overall costs are only first estimates. These estimates are based on data from previous missions, or vendor data. As defined by the challenge, the LCCE is limited to the costs generated by the payloads. An LCCE for the payloads up to Work Breakdown Structure (WBS) level 3 can be seen in table 17.

For external stakeholders, complete time-phased cost plans and schedule range estimates up to WBS level 2 [15], should be ready at Key Decision Point phase B (KDP-B) and high-confidence cost and schedule commitments at KDP-C according to NPR 7120.5F [14]. These estimates in phase B and C of the NASA project life cycle can be calculated using SEER-H by Galorath [48] or, alternatively, cost estimates could be created based on the NAFCOM (NASA Air Force Cost Model) [49].

Name	Ground Penetrating Radar Camera	Lava Tube Exploration Robot	Wind Power Balloon	Mycotecture	Gas Chromato-graph	Deployable Greenhouse	Payloads
WBS Level	Level 3	Level 3	Level 3	Level 3	Level 3	Level 3	Level 2
Cost estimate	$693.500	$174.200.000	$2.107.350.00	$2.104.100	$205.453	$2.106.600	$181.417.003

Table 15: LCCE for payloads up to WBS level 3.

CONCLUSION

It can be argued that all payloads discussed here are well within given requirements of volume and mass, even with a generous margin of error added onto them, with room to spare for other missions. The total required volume should be around 8 cubic meters, with the rover-platform and the greenhouse as the dominating factors. The accumulated weight of all payloads will very likely not exceed 1 ton, leaving a generous 9 tons for the rover-platform.

A possible concern may lie in the technology readiness. We have brought up several payloads with novel and untested ideas, such as the mycotecture, balloons on Mars, snake-robot-drones or power generation utilizing gravity. Here, however, there are either groups already researching and developing these payloads, such as in the case of the mycotecture and snake-robots, or the technologies are very simple, as is the case for the balloon and gravity-power-winched-rover. The development of our rover should therefore not require until 2033 to conclude. Nevertheless, if it had to, it would involve around 600 people working on the project for 10 years. In terms of Full-Time Equivalent (FTE), this corresponds to 1,320,000 FTE required for that project, provided that one person per year represents on average roughly for 220 FTE.

We argue that our approach addresses all of the posed criteria through unique, creative, and valuable solutions. These range from those that explore Mars down from the deepest and most ancient caves, all the way to those soaring high up across its crimson skies. All while keeping a lookout for life that may have been or still inhabits the planet, but also for the life that we will bring and the life that might one day be on Mars. With that we are trying to answer the all-important question whether we can grow and harvest not only food in the form of plants, but also minerals produced by fungi and possibly even create entire habitats using the mycotecture.

We are doing all of that whilst trying to be resourceful and seeking to exploit every potential advantage, including the use of gravity for power generation, and equipping observational balloons with the means of turning wind into electrical energy. To finally tie it all up, these systems all fit-in together to work synergistically with each other to yield the highest benefit for us and our interplanetary future!

REFERENCES

1. Mckay, C. P. The search for life on Mars. Planetary and Interstellar Processes Relevant to the Origins of Life, 263–289 (1997).
2. McKay, C. P., Toon, O. B. & Kasting, J. F. Making mars habitable. Nature 352, 489–496 (1991).
3. Daga, A.W. et al. Lunar and martian lava tube exploration as part of an overall scientific survey in Annual Meeting of the Lunar Exploration Analysis Group 1515 (2009), 15.
4. Zhdanova, N. N., Zakharchenko, V. A., Vember, V. V. & Nakonechnaya, L. T. Fungi from Chernobyl: mycobiota of the inner regions of the containment structures of the damaged nuclear reactor. Mycological Research 104, 1421–1426 (2000).
5. Grant, J. A. et al. The science process for selecting the landing site for the 2020 Mars rover. Planetary and Space Science 164, 106–126 (2018).
6. Jakosky, B. M. & Mellon, M. T. Water on Mars. Physics Today 57, 71–76 (2004).
7. Brown, A. J. et al. Hydrothermal formation of clay-carbonate alteration assemblages in the Nili Fossae region of Mars. Earth and Planetary Science Letters 297, 174–182 (2010).
8. Al Husseini, A. et al. The role of caves and other subsurface habitats in the future exploration of Mars. International Astronautical Congress (2009).
9. Cushing, G., Titus, T. N., Wynne, J. J. & Christensen, P. THEMIS observes possible cave skylights on Mars. Geophysical Research Letters 34 (2007).
10. Greeley, R. The significance of lava tubes and channels in comparative planetology in Origins of Mare Basalts and their Implications for Lunar Evolution 234 (1975), 53.
11. Walden, B. E., Billings, T. L., York, C. L., Gillett, S. L. & Herbert, M. V. Utility of lava tubes on other worlds in Using in situ Resources for Construction of Planetary Outposts (1998), 16.
12. HiRISE spots the mouth of a martian lava tube – NASA mars exploration https://mars.nasa.gov/resources/26349/hirise-spots-the-mouth-of-amartian- lava-tube/.
13. https://www.comparably.com/companies/nasa/salaries.
14. Requirements, N. 7. N. P. NASA Space Flight Program and Project Management Requirements 2021.
15. Terrell, S. M. NASA Work Breakdown Structure (WBS) Handbook tech. rep. (2018).
16. Bell, J. et al. The Mars 2020 perseverance rover mast camera zoom (Mastcam-Z) multispectral, stereoscopic imaging investigation. Space science reviews 217, 1–40 (2021).
17. Mast-mounted camera system (MASTCAM-Z) https://mars.nasa.gov/mars2020/spacecraft/instruments/mastcam-z/.
18. Radar Imager for mars' subsurface exploration (RIMFAX) https://mars.nasa.gov/mars2020/spacecraft/instruments/rimfax/.
19. Grant, J. A., Schutz, A. E. & Campbell, B. A. Ground-penetrating radar as a tool for probing the shallow subsurface of Mars. Journal of Geophysical Research: Planets 108 (2003).
20. Cuadros, J. et al. Mars-rover cameras evaluation of laboratory spectra of Fe-bearing Mars analog samples. Icarus 371, 114704 (2022).
21. Esmaeili, S. et al. Resolution of lava tubes with ground penetrating radar: The TubeX project. Journal of Geophysical Research: Planets 125, e2019JE006138 (2020).
22. Miyamoto, H. et al. Mapping the structure and depth of lava tubes using ground penetrating radar. Geophysical research letters 32 (2005).
23. Lee, P. et al. Lofthellir Lava Tube Ice Cave, Iceland: Subsurface Micro-Glaciers, Rockfalls, Drone Lidar 3D-Mapping, and Implications for the Exploration of Potential Ice-Rich Lava Tubes on the Moon and Mars in 50th Annual Lunar and Planetary Science Conference (2019), 3118.
24. Ford, J., Sharif, K., Jones, H. & Whittaker, W. Lunar Pit Exploration and Mapping via Autonomous Micro-Rover in 2021 IEEE Aerospace Conference (50100) (2021), 1–7.
25. Thangavelautham, J. et al. Flying, hopping Pit-Bots for cave and lava tube exploration on the Moon and Mars. arXiv preprint arXiv:1701.07799 (2017).
26. Kalita, H., Gholap, A. S. & Thangavelautham, J. Dynamics and control of a hopping robot for extreme environment exploration on the Moon and Mars in 2020 IEEE Aerospace Conference (2020), 1–12.

27. Lichtenheldt, R. et al. A Mission Concept For Lava Tube Exploration On Mars And Moon–The DLR Scout Rover in Lunar and planetary Science Conference (2021).
28. Hong, S., Bangunharcana, A., Park, J.-M., Choi, M. & Shin, H.-S. Visual SLAM-based robotic mapping method for planetary construction. Sensors 21, 7715 (2021).
29. Ivie, B. Design of a Controllable Weather Balloon to fly on Mars PhD thesis (Lehigh University, 2017).
30. https://www.esa.int/gsp/ACT/doc/ARI/ARI%20Study%20Report/ACT-RPTHAB-   ARI-16-6101-Fungi_structures.pdf.
31. Hall, L. Mycotecture Off Planet Apr. 2021. https://www.nasa.gov/directorates/spacetech/niac/2021_Phase_I/Mycotecture_Off_Planet/.
32. Rothschild, L. et al. Mycotecture Off Planet: Fungi as A Building Material on The Moon and Mars in Lunar and Planetary Science Conference (2022).
33. Curiosity's SAM https://mars.nasa.gov/msl/spacecraft/instruments/sam/.
34. Curiosity's Mars Methane Mystery Continues https://www.nasa.gov/feature/jpl/curiosity-detects-unusually-highmethane- levels. Accessed: 2019-06-23.
35. Hand, E. et al. Plumes of methane identified on Mars. Nature 455, 1018 (2008).
36. Mysteries in Nili Fossae https://www.esa.int/Science_Exploration/Space_Science/Mars_Express/Mysteries_in_Nili_Fossae.
37. Judd, A. in Atmospheric Methane 280–303 (Springer, 2000).
38. FarmBot Genesis and Genesis XL Description page Kernel Description https://farm.bot/pages/genesis.
39. García de Quevedo Suero (beagqs@gmail.com), B., Capus, L., Faure, N., Heumassej, Y. & Haagh, L. MANGO: MArtiaN GreenhOuse, Proposal for a Planetary Space Mission. Faculty of Aerospace Engineering (2020).
40. Wamelink, G. W., Frissel, J. Y., Krijnen, W. H., Verwoert, M. R. & Goedhart, P. W. Can plants grow on Mars and the moon: a growth experiment on Mars and moon soil simulants. PLoS One 9, e103138 (2014).
41. Paul, A.-L., Elardo, S. M. & Ferl, R. Plants grown in Apollo lunar regolith present stress-associated transcriptomes that inform prospects for lunar exploration. Communications biology 5, 1–9 (2022).
42. Boretti, A. Analysis of the regenerative braking efficiency of a latest electric vehicle tech. rep. (SAE Technical Paper, 2013).
43. Electric hoist winch plastic cable with radio remote control 300kg 12V https://www.seilwinden-direkt.de/navi.php?a=28191lang=eng.
44. Rover "Brains" https://farm.bot/pages/genesis.
45. Rover Temperature Controls https://farm.bot/pages/genesis.
46. Laudon, K. C., Laudon, J. P. & Schoder, D. Wirtschaftsinformatik: Eine Einführung (Pearson Deutschland GmbH, 2010).
47. Gloger, B. Scrum. Informatik-Spektrum 33, 195–200 (2010).
48. Friz, P. D., Hosder, S., Leser, B. B. & Towle, B. C. Blind validation study of parametric cost estimation tool SEER-H for NASA space missions. Acta Astronautica 166, 358–368 (2020).
49. Winn, S. D. & Hamcher, J. W. Nasa/air force cost model: Nafcom in Propulsion for Space Transportation of the 21st Century (2002).

5: THE VULCAN FORGE MISSION

Alden R. O'Cain, Neel S. Shah, and Jason Thai
University of Ontario Institute of Technology
Isaac W.R. Bahler
Dalhousie University
Kai A. Fucile Ladouceur
Confederation College
(kladouceur06@gmail.com)
Samantha B. Chong
University of Waterloo

EXECUTIVE SUMMARY

The attempt to explore and settle Mars has been thoroughly investigated with tens of thousands of people being involved, billions of dollars invested, and years spent. While humanity has been able to explore Mars with the exploration missions it has sent to the planet, few have truly contributed to preparing resources for future settlements. One of the major reasons for this is due to limitations of the technology. In order to truly prepare resources, such as constructing habitation for settlers, additional building material would have to be shipped with the rover, adding on more weight to the payload and money required for the project. However, with recent advances in 3d printing technology, this task is now feasible and does not require the transportation of building material. It utilises the material found on the planet surface and combines it with a basalt composite, creating an organic polymer building material three times stronger than concrete. The Vulcan Forge proposes pairing this technology with a rover sent to the planet through a 10-metric ton payload which would then commence the construction of habitats at site of landing. The printing mechanism would also be housed within a detachable, automated system allowing the rover to pursue independent exploration missions while the habitations are constructed.

OVERVIEW

The exploration and further colonisation of Mars has long since been a dream of humanity, dating all the way back to the 1940s. Since that time, missions have been launched by countries across the world through their respective space agencies in an attempt to make these dreams a reality, however even after decades of attempts by several world powers, and billions of dollars spent, there has yet to be any great headway made in terms of colonising or preparing the planet for future settlers. Critically, this fails one of the primary objectives established by space agencies like NASA,[1] who have narrowed down their goals for Martian missions into four scientific categories:

- Geology
- Astrobiology
- Sample Caching
- Preparation For Humans

While one could argue that these missions all had aspects of them which were preparing for future colonisation by humanity (assessing weather patterns, seismic events, and the internal structure of the planet or transforming oxygen into carbon dioxide), these have not had the impact that one might hope for and really have more of a focus on collecting data and information about the planet than establishing the buildings blocks of a new civilisation. Humanity may be closer to its goal of colonisation than before, but only by a small margin. Now that data and information have been obtained concerning the planet and its environment, it is now the time to truly concentrate efforts on this fourth and very important objective established by NASA and launch a mission which will determine whether or not humans could settle the planet. This report will go into the details of why constructing habitations will be the first key step in truly preparing for human arrival on the Red Planet, and how this has become a much more viable option now than ever before. While true, before recent scientific breakthroughs, the idea of constructing habitations was a far-fetched and truly expensive endeavor as trying to transport a rover to the surface alone was a task in itself, let alone adding onto this the weight of building material for homes. Nevertheless, science has found the solution to this perplexing problem by offering the gift of 3d printed habitations which utilise the materials found on the surface of the planet as construction material.

Objectives

(a) Close-up

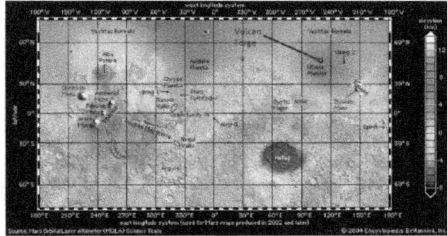

(b) Relative Position

Figure 1: Views of proposed landing site

When defining parameters of success for a mission it is essential to consider what the objectives of the mission are. In this case, the primary overarching goal of the Vulcan Forge is to prepare for human settlement on Mars. To accomplish such an enormous goal this mission breaks the task down into smaller components aiming to design technologies to 3D print human habitats, Materials Processing, Power production, subterranean exploration and sustainable resource procurement. Each component of the mission aims to prove an existing technology in the field elevating its NASA technology Readiness Level and demonstrating the equipment in the field. Additionally, the subterranean exploration and additional analysis equipment will expand the growing body of knowledge about Mars, particularly about what precisely exists below the surface, largely unknown. When taken together then the objective is fundamentally to test all major technologies required for human survival on the surface that lack field tests while simultaneously collecting information on little-known regions of the Red Planet.

Site Selection

When considering a landing site, it is essential that the location is suitable for the objectives of the mission meaning abundant resources and unexplored territory. As such the Vulcan forge Mission will land in Utopia Planitia a largely unanalyzed region near the Viking 2 lander.[2] A rover in the area, particularly one equipped with a means for subsurface exploration, will allow the deep rifts and volcanically rich area to be reviewed in depth. An alternative explanation for the intriguing imagery sent back from the lander and satellite reconnaissance is past sediment settling and mud eruption.[3] If that is the case the area has a high likelihood of harbouring signs of past life making it extremely scientifically interesting. The specific site of interest we would aim to target is a relatively new crater on the surface both for geological exploration and the prevalence of materials for construction testing.[4] Landing in such an area gives the Vulcan Forge terrain that can be navigated by a rover, interesting sites for scientific exploration and the needed materials to test the primary objective of the mission.

THE DESIGNS

Primary Drone/Rover

The primary rover in the Vulcan Forge fleet follows the ordinary six-wheel drivetrain implemented in Curiosity and Perseverance, however scaled up slightly to handle the additional weight from added instrumentation and subsystems that do not exist on previous Martian rovers. Instrumentation from previous missions to Mars as well as other research can be reused; for example, WATSON/MAHLI (topographic sensors), RIMFAX (subsurface radar), MEDA (weather station), Mastcam-Z (high resolution imagery) can all be utilized as is or with improvements as of a result of a new and improved version. The rover design can be seen in Figures 2 – 4.

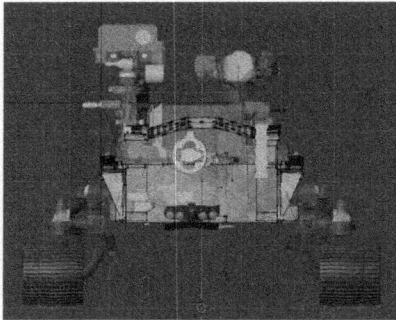 

(a) Front   (b) Back

Figure 2: Front and back views of tiles application in rover

(a) Left   (b) Right

Figure 3: Left and right views of tiles application in rover

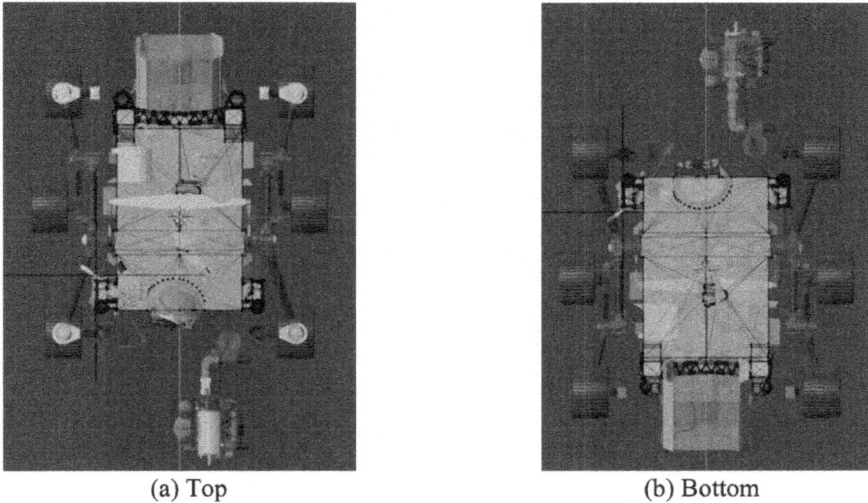

(a) Top   (b) Bottom

Figure 4: Top and bottom views of tiles application in rover

Drivetrain & Chassis

Part of the chassis of the primary rover would also feature a hitch system for towing the printing drone which by itself is extremely limited in power in its drivetrain and only exists for the purposes of printing. For relocations and longer journeys the printing drone would attach using a hitch that resembles a fifth wheel that can also provide power and transfer data to the main rover for the purposes of analysis and or transmission back to Earth.

Sensor Suite

Because Vulcan Forge would need to operate autonomously at times, it's sensor suite would have to be extensive. Omnidirectional imagery can be provided by an array of eight navigational/hazard cameras,[5] Mastcam-Z/SuperCam to capture imagery of nearby targets, WATSON/MAHLI (topographic sensors),[6] RIMFAX (subsurface radar), atmospheric temperature, pressure and winds can be measured with a MEDA (weather station) device. Figure 5 shows a schematic for an omni directional camera, like those found on the Vulcan Forge rover - would include a rotating platform with multiple cameras mounted on it. These cameras would be able to capture images in all directions, allowing the rover to survey its surroundings and gather data on the Martian landscape. The platform would be driven by a motor and controlled by a computer system, which would coordinate the movement of the cameras and the collection of images. Figure 6 pictures a weather station, similar in nature to the one mounted on the Vulcan Forge rover, capable of capturing environmental metrics on the Martian surface. Using a thermometer for measuring temperature, an anemometer to gather wind speed, and a barometer for atmospheric pressure, however additional sensors and metrics would likely be included. These sensors would be connected to a data logger or computer system that could record and transmit the collected data. Figure 7 shows LIDAR imagery can be used to locate underground caves by emitting pulses of laser light and measuring the time it takes for the pulses to return after bouncing off the surface of the ground. By analyzing the return times and the intensity of the pulses, it is possible to build up a detailed 3D map of the terrain, including any underground cavities or structures. Figure 8 shows a ground penetrating radar image used to locate utilities underground on Earth through use of a specialized radar system that can transmit and receive signals through the ground. By analyzing the reflected signals, it is possible to build up a detailed map of the subsurface features, including any underground pipes, cables, or other utilities. This type of imaging can be useful for utility companies or construction firms, as it can help them to avoid damaging or disrupting underground infrastructure.

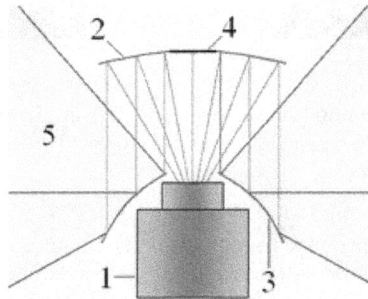

Figure 5: Diagram of omni directional camera on Vulcan Forge

Figure 6: Weather station

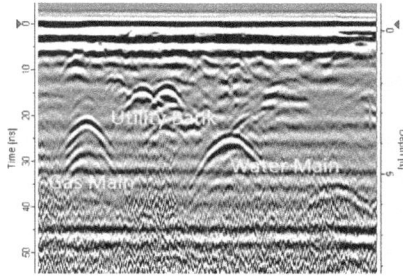

Figure 7: Diagram of LIDAR system on Vulcan Forge

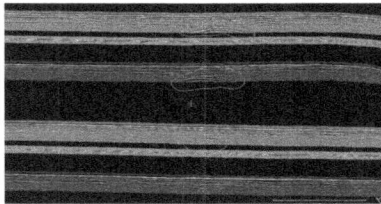

Figure 8: Ground penetrating radar image

Drilling and Material Analysis

Inspired by the work of Quinn Morley and Tom Bowen we have included their concept of boring robots for drilling the Martian ground for the purposes of obtaining soil and ice samples.7 The boring robots (shown in Figure 9) could drive up and down holes that they drill and take turns to drill to deeper depths even for the purposes of finding submartian water reserves8 since Orosei et al. 2018 suggested mainly, that a subglacial liquid water lake is hypothesized to exist. The boring robots would also take small samples for analysis by scientific equipment on the main rover.9 While stowed they would exist in a large carrier similar to that of a magazine on a firearm, but entry in the rear, to sequence a returning boring robot to the back to be recharged while another is deployed. With a borehole created it would also be possible to send instrumentation down into the hole on the chassis and drivetrain of a boring robot, instrumentation such as laser spectrometers, and laser microscopes.

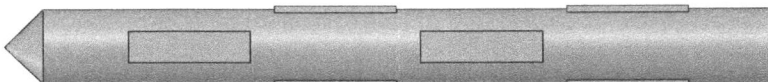

Figure 9: A Boring Robot

Movable Feast Machines (MFM)

The standard that has been established for computing devices relies heavily on a deterministic system, with any particular input yielding a consistent output.10 While accurate and reliable by itself, the standard is limited: all computations must happen in the CPU in order to interact with applications and programs, and without it, it would become unusable. Other issues regarding the architecture require the user to completely reset the program when a specific hardware or software error occurs, as well as being unable to expand outside of its environment.

An example of the problems current rover software systems face is NASA's Curiosity Rover. On November 29th, 2011, a reset

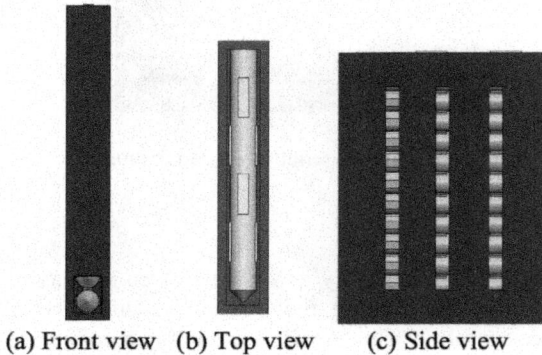

(a) Front view   (b) Top view   (c) Side view

Figure 10: Boring Robot Silo: Front, Top and Side Views

occurred three days after launch, during use of the craft's star scanner. The cause had been identified as a previously unknown design idiosyncrasy in the memory management unit of the Mars Science Laboratory computer processor. In rare sets of circumstances unique to how this mission used the processor, cache access errors could occur, resulting in instructions not being executed properly.11

The rover design had 2 central processors, the main processor, called the pilot (A-side), and the co-pilot (B-side) as a backup.12 Both equipped with shielding from space radiation, 2 gigabytes of memory and 256 megabytes of dynamic random-access memory (DRAM).13 The computer was located inside the Rover Computer Element (RCE) that interacted with all the rover components.13

The pilot refused to give control to the co-pilot while experiencing a memory problem. Luckily, the co-pilot was able to take control, however had it failed to perform the task, it would have meant the mission was lost and a loss of $2.5 billion.14, 15

There are known automated testing frameworks used to prevent software failures such as CodeSonar,16 however this adds an extra layer of complexity to the stack and cost.

Movable Feast Machines (MFM) solve the problems associated with the standard that is used in existing rover systems' today (such as restrictions on memory, scalability and use of a central processing unit (CPU)) while providing a flexible architecture similar to living organisms.10 The devices are shaped like tiles that are able to expand using the idea of indefinitely scalable architecture which supports open-ended computational growth without re-engineering.10 This kind of scalability rejects all internal limits, so machine size is limited only by external costs such as real estate, materials and construction, and power and cooling.10 Table 1 compares and contrasts the Curiosity Rover solution13 and the MFM solution.10

One of the current flaws with the system is that it is currently not recommended to use for higher level tasks such as monitoring sensor data, temperature and any important instruments because the software receives those messages and compares what is expected and then makes copies of the computations until it fills up all the hardware available. This then makes the entire system redundant and therefore difficult to notice errors since the neighbouring tiles can ignore, fix, or erase and re-establish locally, and branch out to other tiles globally. This leads to a suggestion of a hybrid of the traditional and distributed architecture.

Table 1: A table comparing the Curiosity Rover solution and the MFM solution.

Curiosity	MFM	Potential Solutions
Idiosyncrasy Memory Design	Processors and Memory is distributed	Tiles being categorized/labelled in their respective areas within the rover by utilizing the caching and locking model to perform local computations can help properly distribute where separate components may need support to prevent a system- wide shutdown.
Synchronous Communication (Pilot and Co-pilot)	Asynchronous Communication (Spatial Computing)	Reproducing and distributing (updating) to other available tiles where necessary. Updates don't have to

		synchronize to prevent a case where a timing error can cause a system-wide disaster.
Central Processing Unit	No Central Processing Unit	Each tile has their own Node 'atoms' (bits) to communicate with each other (reading and writing), therefore not relying on a single CPU architecture but rather individual tiles (with implemented correction techniques pre-programmed). It is beneficial to have separate components supporting each other acting as backups.

The traditional CPU and RAM can deal with high level tasks while the MFM can act as a way to monitor and keep the system running. Furthermore, having models of the surrounding environment can possibly make predictions to prevent fatal errors and update or assign tasks from other rovers to take over if it occurs, similar to traditional system controls except it is spread out spatially.

In conclusion, this concept applies to all or most of NASA's telerobotic systems and their instruments, following hybrid modelling. To further elaborate on the concept, see the onion diagram and flowchart in Figure 11.

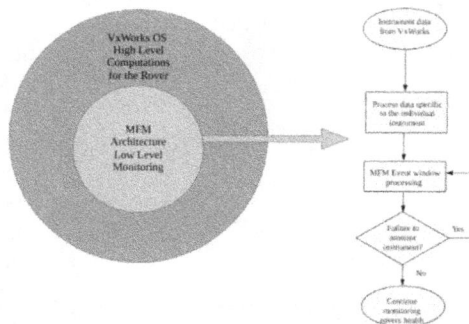

Figure 11: Onion diagram and flowchart of MFM

The code behind MFM runs on ULAM, an object-oriented procedural programming language.17 The ULAM compiler is based on C++ and the language itself is applicable for embedded programming, especially in VxWorks, which is the software the Perseverance rover uses. ULAM is then passed to SPLAT (Spatial Programming Language ASCII Text) to further enforce the idea of asynchronous cellular automata. It displays the information of the individual atom and which direction it is facing.

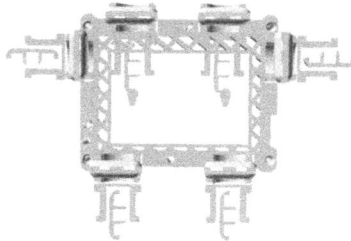

Figure 12: Assembling the MFM tiles along with its connectors. Each casing has etched directions of: North, South, East and West

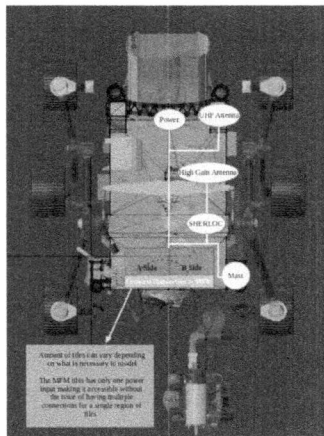

Figure 13: Application of T2 tiles. A follow up on the onion diagram and flowchart, representing data from the rover's specific instruments passing to computers A or B, then passing through the MFM to monitor its health.

Power System

The power generating capabilities of a mission ultimately define what the mission can accomplish as a whole, the greater the power output capabilities the more processing and analysis that can be accomplished. As such optimizing power production for the nature of the mission becomes critical. Given the primary objectives of the Vulcan forge initiative, a consistent, and exceedingly high wattage power sup- ply is needed to power both the rover and the equipment it possesses. Processing 3D printing materials successfully creating structures will undeniably require more power than any previous NASA mission by orders of magnitude. When considering methods of generating power for a Martian mission it is essential to consider the planetary structure and makeup of Mars as a whole. There are numerous factors which make conventional power generation impossible.

Atmospheric density is low, approximately 1% of Earth's meaning that conventional wind power generation is significantly less effective (Table 2). Furthermore, large the commonality of large windstorms means that any turbine would have to be built to withstand high force making them more robust than needed the majority of the time, decreasing power generation. When considering geothermal power generation on mars we lack the sub- terranean research and probing necessary to confirm the reliable presence of power. Taken together these factors make both geothermal and wind power poor candidates for an initial mission.

That being said, the subterranean exploration component of the Vulcan Forge initiative will conduct the reconnaissance needed for geothermal power to be an option preparing a future settlement for another power option. Due to the surface temperature, a lack of liquid water invalidates hydroelectric options entirely, though seemingly simple a necessary consideration as depending on the atmosphere liquids still often flow on the surface of planets. This leaves solar and Nuclear as near-exclusive options for powering our mission in the long term. Due to the decreased solar efficiency on a planet farther from the sun, dust storms lowering viability even further and the difficulty in moving the vast amounts of solar panels that would be needed to provide suffei9cnt power for the industrial rover Nuclear was deemed a superior option.

After nuclear was determined to be the most viable solution it becomes only a matter of picking the technology best suited to a test that would likely benefit future colonists. With the hope of increasing mission simplicity, a new generation of space nuclear reactors was selected collectively known as the Kilopower system. These reactors use Stirling engines to convert heat from liquid sodium circulated around decaying uranium to electrical energy. Using the proposed size iteration of a 5kWt rector the power needs of Vulcan Forge would be more than met. The reactor uses 93% enriched Uranium undergoing fission and needs minimal shielding as there are no living organisms onboard the mission. This mission would provide an environmental test for a technology already proven in the lab and prepare the way for a larger reactor with increased shielding to protect astronauts in the future. The proposed design and weight constraints can be seen in Figures 14 - 15. The simplicity of the reactor and reliability and easy reconfigurability of the design make integrating it here an ideal candidate. The authors argue it could easily be reshaped to suit any mission and that its low cost is economically feasible as well as technically viable. Additionally, due to the constraints of the competition, the reactor is compact, and shielding can fold to decrease volumetric capacity during launch allowing for easy transport.

(a) Power System   (b) Power System Layout

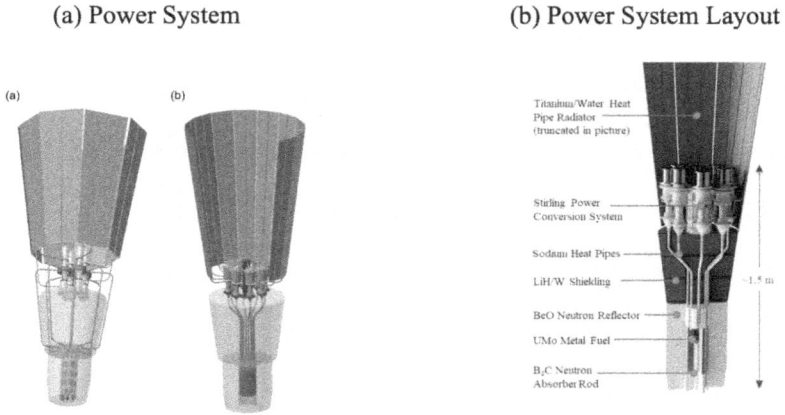

Figure 14: Kilopower Reactor System

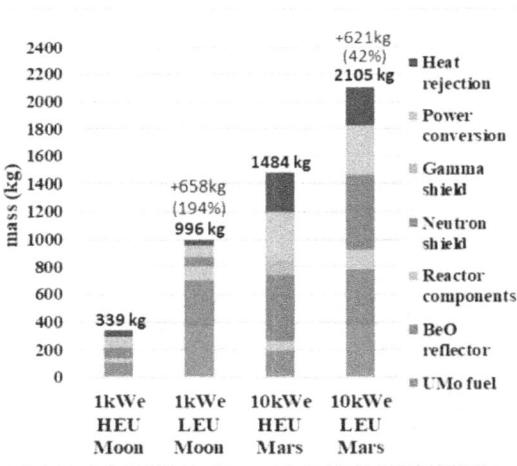

Figure 15: Space Kilopower System - Relaxed Radiation Dose

Table 2: Comparison of Earth and Mars

Planetary Property	Earth Condition	Earth Condition
Atmospheric Pressure (Surface Level)	1013 millibars	6-7 millibars
Atmospheric Makeup	95% - CO_2 1.9% - N 1.9% - Ar Trace Gases	20.9% - O 78.1% - N 0.93% - Ar Trace Gases
Gravity	9.81 m/s2	3.71 m/s2
Geological Activity	Well Quantified	Significant Conflicting
Temperature Variation	Minimum -88 oC Max 58 oC	Min -140 oC Max 30 oC
Day Length	24 hours	24 hours, 37 minutes
Planetary Magnetic Field	Yes	No
Mass	5.97 x 1024 kg	6.42 x 1023 kg
Average Density	5514 kg/m3	3933 kg/m3
Surface Area	5.10 x 108 km2	1.44 x 108 km2

Robotic 3D Printer Arm

Additive manufacturing allows digital models to be created in the real world. A major portion of additive manufacturing comes from 3D printing. This manufacturing method allows complicated designs to be quickly and efficiently built. In additive manufacturing, structures can be built before humans physically come to mars, meaning that an entire habitat could be built, set up and tested without the risk of human life. These 3D printers would be able to build and manufacture the structures necessary for entire communities with minimal human intervention.

The main 3D printing mechanism is based on a robotic arm 3D printer. Robotic arms have two main benefits. They have a higher capacity for detail and are more portable. Our proposed robotic arm will be placed on top of the rover, allowing it to move. This 3D printer is based on the design by Steven Middleton for his pocket 3D printer project.[18] This printer was designed to be small enough for an individual to carry around in their pocket. Using some of the principles from this design, we could create a compact arm for this 3D printer. To expand this project's build capacity, the mechanism from design 349 from *507 Mechanical Movements* by Henry T. Brown was adapted and utilized.

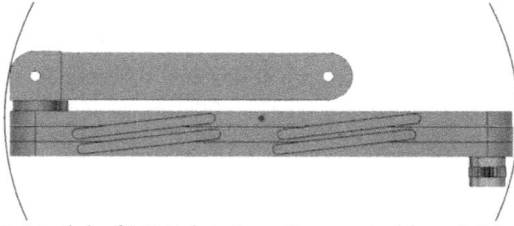

Figure 16: Model of 3D Print Arm Compacted in a 5-Meter Radius

As seen in Figure 16, the mechanism from design 349 allows the arm's height to expand from just under 5 meters to about 10 meters, double the original height. Overall, when the rover is stationary, the 3D printer arm has a build volume of approximately 550 m³. In addition, the rover is also mobile, allowing the build volume to significantly increase in both the x and the y planes. (Note that in 3D printing, the Z plane is traditionally the vertical plane)

Extrusion

Most commonly found desktop 3D printers use either a Bowden extruder or a direct drive extruder. Due to the nature of the material which this printer will be using, these extrusion methods are ineffective. The material this printer will be extruding will be a concrete-like compound using materials native to Mars. The compound will be heated and deposited into the system in a liquid form. In order to extrude the material, this system will use a boom concrete pump. This is possible due to the similarities in properties between the two materials. The Martian concrete will be brought to the 3D printer arm via a modified boom concrete pump. The image in Figure 18 displays an example of what this system could look like.

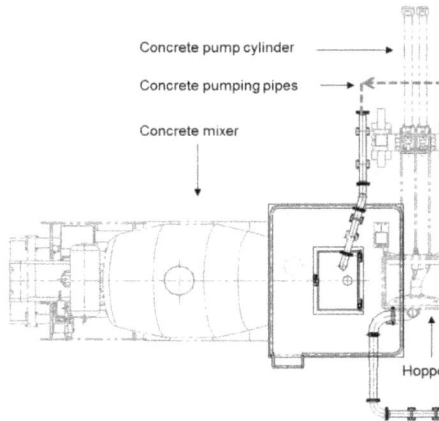

Figure 17: Diagram of boom concrete pump

PLA and Polyethylene

As the Vulcan forge is fundamentally a prototype mission designed to test theoretical technology the mission will aim to analyse the effectiveness of several different materials proposed in past habitat designs. When selecting materials to test strong preference was placed on materials that can be sustainable procured on Mars via on site resources collection limiting material options. The material additionally needs to have properties conducive to habitat construction and 3-D printing techniques. Martian regolith is primarily composed of silicon dioxide and ferric oxide with further large components of aluminum oxide, calcium oxide, and sulfur oxide. Additionally, readily available at the surface are large deposits of basalt as this is one of the primary components of Mar's rocky surface along with water and Co2 ice in addition to the abundant atmospheric Co2. Existing analysis of different materials mainly propose combinations of materials available on Mars.

When 3D printing ideally long polymer chains of carbon are desired as they are easier to mold but form relatively strong structures. As such two frequently proposed polymers are Polylactic acid and Polyethylene seen in Figure 18. While Stable and ideal to print using well tested technology from earth both polymers lack the strength necessary for long term habitation structures. Ideally then both will be reinforced with materials taken form the surface of the planet.

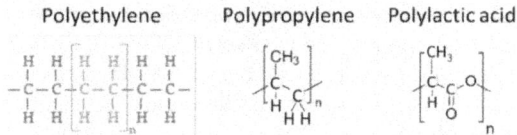

Figure 18: Chemicals used in building construction

Supporting Polymers - Basalt and Sulfur

When considering material to support each polymer forms strong materials that are more difficult structure come to mind. Sulphur concrete extracted from the surface of the planet for instance has proven to be exceedingly strong and ideal for Martian conditions but difficult to work with. Basalt, an abundant mineral on the Martian surface for instance could be collected and turned into a fibre using commercially available processes. When mixed into standard polymers, this greatly increases their ambient strength and resiliency to the Martian environment. Sulfur concrete, a mixture of standard Martian regolith and liquid sulfur, could even fill polymer printed molds providing an internal framework to quickly construct the outer layer. Together, this use of material maximises the use of Martian materials while simultaneously maintaining stability and strength in the designs being tested.

Collection and Processing

When considering how to collect and process Martian material an easily added conveyor system on the bottom of the rover could collect rocks and dust as it moved. Following that a brief spin in a small centrifuge could separate different compounds after which sulfur and basalt could be separately processed for materials testing, both require grinding and sulphur additional heating; all of which can be accomplished with modified standard industry equipment integrated into the rover. Polyethylene or a similar synthetic could be synthesized using carbon capture from the air and water collected using a microwave system removing water from the previously spun Martian regolith. Together this should allow for the sustainable production of all printing materials which can then be piped to the printing arm in heated tubing to ensure viscosity is maintained. The only, material which cannot be produced infinitely is PLA of which a stock must be brought for testing from Earth.

Materials

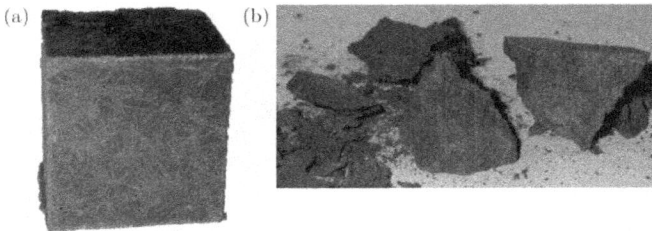

Figure 19: Sulphur and Basalt based materials

The amount of material used will be calculated based on the surface area of each part. It will be assumed that 20% will be structural using 3 mm aluminum. The other 80% will comprise of 0.5mm aluminum. For this structure, the main building material will be aluminum alloy 7075 due to its high strength-to-weight ratio. 7075 has a slightly higher density than pure aluminum at $2.81g/cm^2$.[19] Figures 19 and 20 demonstrate hypothetical concretes for us in the construction of Martian habitats.

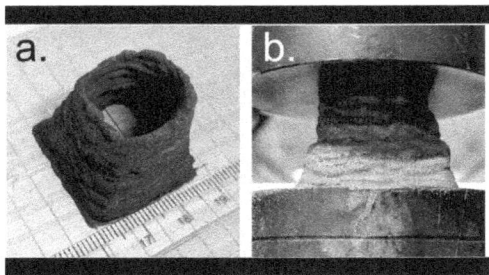

Figure 20: Demonstration of properties of Sulphur and Basalt structure

Habitation Design

(a) SEArch+/Apis Cor Mars Ice House[20]   (b) Zopherus Beetle Habitat[21]

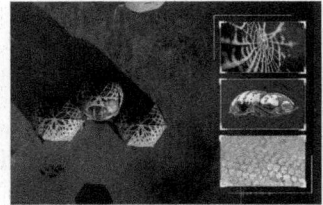

Figure 21: Artist Renditions of Habitats

One of NASA's recent Centennial Challenges was based on using additive manufacturing to build habitats for future astronauts. We will use the results from those reports as a basis for our proposed base, although it is essential to note that due to the nature of our 3D printer's robotic arm, it can print most, if not all, of these designs. The proposal for the habitation section of this report is that the top two proposals won in the Centennial Challenge were adapted and used as inspiration to create a proposed habitat with the positive aspects of each of these habitats. The first-place winner was SEArch+/Apis Cor. They proposed using a thin layer of ice to protect against the dangerous radiation on the surface of Mars. They propose spraying water through a heated nozzle onto an inflated surface. The water would then freeze into ice, providing a radiation shield which allows the wavelengths of visible light to enter the structure. Figure 21 shows a visual rendering of this. The second-place winner was Zopherus, which based the structure of their habitat on that of a Zopherus beetle. Their structure was also hexagonal to allow the structures to be modular. The habitat proposed in this build will utilize the idea of using ice as a radiation shield and a hexagonal shape to increase the modularity. The base square footage of each module is just over 100 square feet, slightly smaller than the average bedroom. Each module contains three stories with stairs connecting them, providing the total square footage of 300 square feet.

The habitat will be constructed in three stages, two of which can be done autonomously before a manned mission. The first stage is the main structural building of the habitat using Martian concrete. The second stage is to cover the outside of the structure in ice, allowing for additional protection from the radiation present on the surface of Mars. The final stage will be the interior of the habitat, this portion of the habitat will require additional materials from Earth and, as such, will not be part of this mission as it is entirely autonomous.

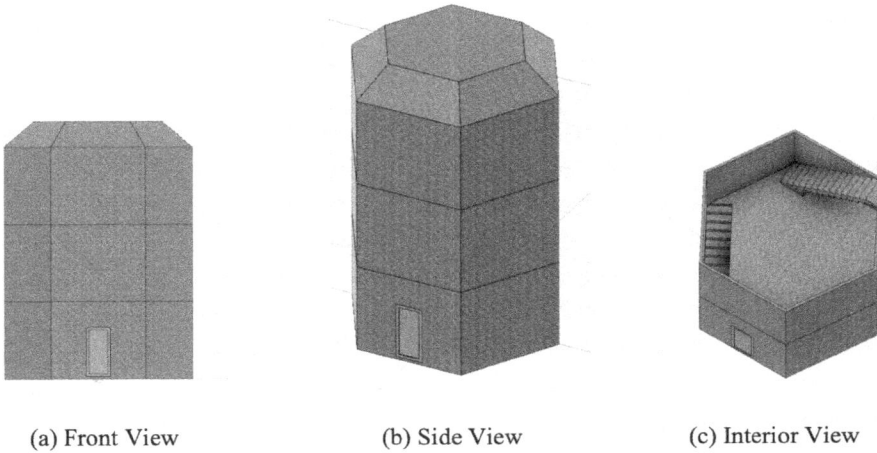

| (a) Front View | (b) Side View | (c) Interior View |

Figure 22: Views of Modular Habitats

THE PLAN

Startup Sequence

During the first thirty to seventy-five days the focus of the mission will be on Vulcan Forge validating itself. It will be during this time that the different systems and instrumentation in the convoy will be tested for the first time on the Mars.[22] Everything from the drivetrain to the extrusion of the printing arm must be confirmed working and operational before the mission truly begins and the first flight of the exploration MAV takes place. In the event of equipment arriving damaged or nonfunctional then the operational planning of the mission will likely need to be adjusted going forward working around faulty equipment but still attempting to work towards completing the mission and achieving the science and exploratory goals. After the majority of the instrumentation is confirmed to be working Vulcan Forge will be able to test its drive train by moving for the first time on the surface of the red planet, during this time it will also be able to test for the first time the microwave-based water acquisition system while in motion since the yields from stationary testing with have diminishing returns in terms of captured water.

Once locomotion is confirmed a test drilling site may be determined for the purpose of deploying a boring robot and testing its systems as well. This would be a conservative test that does not require the entire capacity of a boring robot's battery reserves. Ideally all systems on the first deployed boring robot are confirmed to be working and it is able to be restored inside magazine of the Vulcan Forge's main rover. This test will have to be repeated for every boring robot to confirm that they are reliable before future use.

CONCLUSION

To conclude this report, Vulcan Forge will be assessed by the five different categories provided by the TMEDC:

- Engineering Design
- Scientific Return
- Exploration Preparation
- Cost
- Schedule

Engineering Design

In terms of engineering design, the design is absolutely feasible and implementable. The technology's utilised by Vulcan Forge are proven and are either in the prototype or implementation stage of their development. To effectively measure the missions technology feasibility an estimate was made using NASA's Technology Readiness Level (TRL), the system ranks existing technologies according to reliability and past testing. The research presented here consists entirely of ideas achieving a TRL of 4-5 meaning the only way to further improve upon the ranking is to conduct in environment tests on location and further improve technology.

Furthermore, it falls within the weight and volume metrics laid out by the competition, weighing just over 2 900kg and having a total volume of $23.4m^3$.

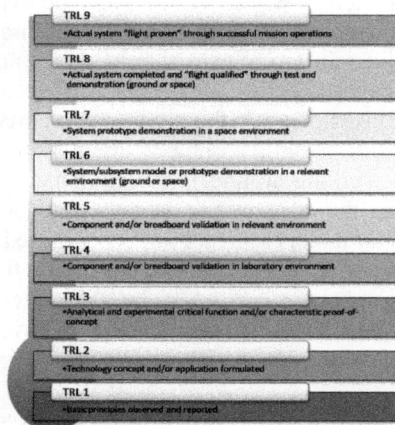

Figure 23: Technology Readiness Level (TRL)

Science Return and Exploration Preparation

When assessing the value a mission provides the im- pact on all parts of the scientific ecosystems must be consider. The Vulcan forge initiative aims to provide a concrete test for numerous proposals designed from earth. Without field testing the value and feasibility of technology built for human settlement on Mars is severely limited. As outlined here the Rover and accompanying equipment will be capable of conducting subterranean geological analysis; further surface exploration and charting; in-situ resource collection and processing of valuable elements like Water, Carbon and Sulfur; 3D print testing, a T2 Tile computing system and a next generation nuclear Kilopower Generation system all integrated into one rover. When taken together the scientific return such a mission will lay the groundwork for future human habitation. Geological exploration will examine materials and structures available below the surface and examine the viability of geothermal. Resource extraction from ice and regolith will test the feasibility of humans purifying organic molecules necessary for long term survival. The T2 tile will test living systems principles in onboard computing to increase reliability and scalability. Every component of the Vulcan Forge advance humanity's collective understanding of how humans can adapt to surviving in a long term space environment. The scientific impact of the testing front running space habitation technologies now will allow for massive leaps forward in the construction of future crewed missions both to Mars and Beyond.

Cost

Part of determining the feasibility of a project is in the project's cost. It is important to note that for a mission of this scale, it is next to impossible to determine the exact cost as many unforeseen issues may present themselves. However, the cost estimated must be reasonable, accounting for some unforeseen issues, but not everything as that is unreasonable.

Overall, it is estimated that the entire arm assembly and related components will cost around 30 million USD. In addition to this the 3D printer arm requires materials to print, it is estimated that the cost of manufacturing and developing these materials is 30 million USD.

Based on the cost of NASA's Perseverance Mission in 2020, we can estimate that the base cost, including the rover, would be about 250 million USD. This covers the base cost for the operations and analysis for two years. In addition, it is estimated that the Boring Robots would cost around 75 million USD to build and implement for this mission. It is estimated that the T2 Tile program would cost around 40 million USD to implement and maintain. The nuclear reactor proposed for this mission is estimated to cost around 20 million USD to prototype fully. Adding the possible failures and other probable issues, it can be estimated that the reactor would cost about 50 million USD to build and fully implement into the rover.

Adding these costs, we propose that this mission will cost 475 million USD, not including the cost of the actual launch. Mars Perseverance cost $2.55 billion minus the launch cost, making Vulcan Forge significantly cheaper while being of a similar magnitude.

The proportion of the cost taken up by each component of the mission can be seen in Figure 23.

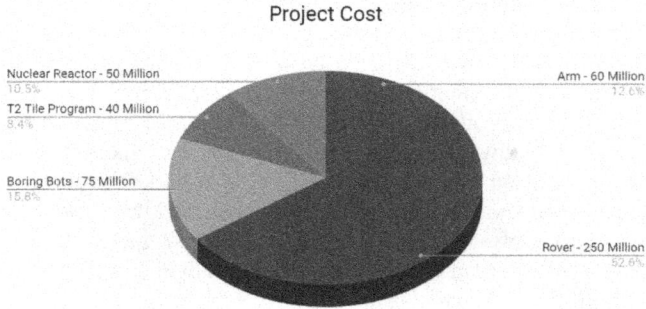

Project Cost

Nuclear Reactor - 50 Million
10.5%

T2 Tile Program - 40 Million
8.4%

Boring Bots - 75 Million
15.8%

Arm - 60 Million
12.6%

Rover - 250 Million
52.6%

Figure 24: Visual breakdown of project cost

Schedule

To determine when this mission could be made, it is important to determine the windows of opportunity when a Hohmann transfer can be successfully completed, wherein Earth's perihelion orbit would align with Mar's aphelion orbit (as shown in Figure 24).23

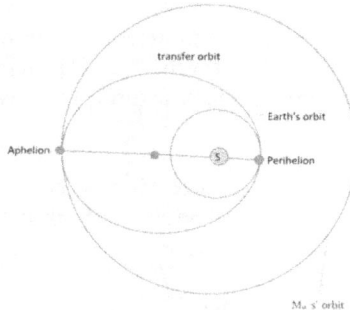

transfer orbit

Earth's orbit

Aphelion   Perihelion

Mars' orbit

Figure 25: Hohmann Transfer Launch to Mars

By using the most recent mission launched to Mars, Perseverance, as an example for what the launch window would look like, and restraining the list of dates to those within a few years before and after the selected date of 2033, this would limit the launch dates to approximately: February 7*th*, 2028; December 25*th*, 2029; November 11*th*, 2031; September 29*th*, 2033; August 16*th*, 2035; and finally July 3*rd*, 2037.

To determine these dates, as previously mentioned, the Perseverance Mars mission launch was studied. From this, it was discovered that the trip only took approximately seven months to reach the Martian surface.[24]

This would have meant that Mars would have been 70 degrees ahead of Earth in its orbit (as Mars moves approximately 0.524 degrees per day * approximately 210 days to reach Mars, which would result in 110 degrees travelled by Mars after 210 days. This would then be subtracted from Mars' position when its orbit would be in the right position, which would be 180 degrees from the vernal equinox, 70 degrees in relation to Earth's orbit is obtained).[23] After calculating the position relative Earth Mars would have to be in, other dates could then be computed by adding 687 days to the launch date (which is approximately the number of days it requires for Mars to complete one full revolution around the sun) (see Figure 25 for graphical representations of the calculations).[23]

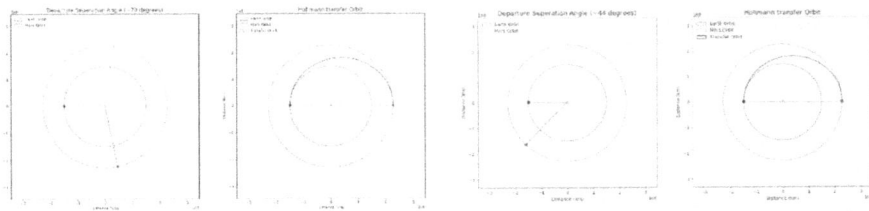

(a) Average   (b) Perseverance Estimated

Figure 26: Hohmann Transfer Launch Calculations to Mars

It should be noted that Perseverance was an anomaly as the trip normally takes closer to 9 months or 259 days[23] and thus this would change the calculations with the departure separation angle being closer to 44 deg rather than 70 deg. This would mean the payload would have to leave the planet sooner and alter the dates which have an opening. However, since the paper has thus far based itself upon the previous Perseverance mission, it will also utilise the associated Hohmann calculations.

After taking these dates into account, it is then important to factor in the preparation time required before launching the payload. Using the Mars 2020 mission as a basis for the amount of time required to prepare for a launch, mission would have to be somewhere around seven to eight years from now (as the first meetings held to discuss and prepare for the mission began as early as December 3*rd*, 2012).[25]

This would eliminate any dates before 2030. In addition to this, missions would have to also fall within a date outside of the Martian dust storm season which usually occurs during the summer every five-and-a-half Earth years, and can last up to a few months.[26] The last storm was in January 2022, so it would be safe to assume that a launch date at least several years after 2027 would be ideal.[27]

Finally, it is important to note that a majority of the technology used in this proposed mission are either technologies reused from previous missions to Mars or have been developed to the point where it would be feasible to implement them in a mission to Mars after a several more years of fine-tuning. Thus, the predicted deadline for launch would be November 11, 2031 and hence, the rover should be expected to actually land on the planet by sometime around June 11, 2032.

The 3D-printed habitation construction that is possible with Vulcan Forge pushes humanity even closer its goal of colonizing Mars.

ACKNOWLEDGEMENTS

The authors acknowledge the assistance in ideation from Dr Dave Ackley and Dr Peiying J. Tsai as well as ideation and technical support from Russell Frost, Benjamin Fedoruk and Kyla Pun.

REFERENCES

1 NASA. Objectives, 2020.

2 University of Arizona, NASA, and JPL-Caltech. Utopia Planitia.

3 NASA, JPL, and Arizona State University. Utopia Planitia.

4 Kenneth L. Tanaka, James A. Skinner Jr., and Trent M. Hare. Geologic map of the northern plains of Mars. *USGS*, 2005.

5 NASA. Curiosity's cameras. https://mars.nasa.gov/resources/20131/curiosityscameras/, 2013.

6 NASA. Mars hand lens imager (mahli). https://ntrs.nasa.gov/api/citations/20050166874/downloads/20050166874.pdf, 2005.

7 PlanetEnterprises. Borebots: Tetherless deep drilling into the Mars south polar layered deposits. https://www.planet.enterprises/post/borebots, 2021.

8 PlanetEnterprises. Borebots niac presentation 1080p. https://www.youtube.com/watch?v=UL9BN0kimYw&ab_channel=QuinnMorley, 2021.

9 PlanetEnterprises. Borebots: Unlocking subglacial lake access - morley/bowen

- 2021 mars society virtual convention. https://www.youtube.com/watch?v=QpJaufQcgPM&ab_channel=TheMarsSociety, 2021.

10 David H Ackley, Daniel C Cannon, and Lance R Williams. A movable architecture for robust spatial computing. *The Computer Journal*, 56(12):1450–1468, 2013.

11 Guy Webster. Spacecraft computer issue resolved, Feb 2012.

12 Andrew Good. Curiosity on the move again, Nov 2018.

13 NASA. Brains of mars curiosity rover.

14 Amy Svitak. Cost of NASA's next mars rover hits nearly 2.5 billion, Feb 2011.

15 Mike Wall. A glitch nearly killed NASA's curiosity rover after 6 months on mars, May 2017.

16 GrammaTech. Curiosity's software upgrades study GrammaTech.

17 Dave Ackley. Living computation foundation programming soft alife with splat and ulam.

18 Brian Krassenstein. Pocket3dprinter, unique, foldable, pocket size 3d printer, launches on Indiegogo for 299. https://3dprint.com/18240/pocket3dprinter/, 2014.

19 MatWeb. Aluminum 7075-t6; 7075-t651. https://www.matweb.com/search/DataSheet.aspx?MatGUID=4f19a42be94546b686bbf43f79c51b7d&ckck=1.

20 SEArch+. Mars ice house. https://www.spacexarch.com/mars-ice-house.

21 Zopherus. Zopherus design. https://www.zopherus.design/.

22 MoreInput. Mer timeline English 2014. https://commons.wikimedia.org/wiki/File:MER_Timeline_English_2014.png, 2015.

23 NASA. Let's go to mars! calculating launch windows. https://www.jpl.nasa.gov/edu/teach/activity/lets-go-to-mars-calculating-launch-windows/ #breadcrumb-accordion-1, 2016.

24 NASA. 2020 mission perseverance rover | mission timeline: Cruise. https://mars.nasa.gov/mars2020/timeline/cruise/#:~:text=The%20spacecraft%20departs%20Earth%20at,miles%20(480%20million%20kilometers)., 2020.

25 J.F. Mustard, M. Adler, A. Allwood, D.S Bass, D.W. Beaty, J.F. Bell III, W.B. Brinckerhoff, M. Carr, D.J. Des Marais, B. Drake, K.S. Edgett, J. Eigenbrode,

L.T. Elkins-Tanton, J.A. Grant, S.M. Milkovich, D. Ming, C. Moore, S. Murchie, T.C. Onstott, S.W. Ruff, M.A. Sephton, A. Steele, and A. Treiman. Report of the Mars 2020 science definition team. September 2015.

26 Kathryn Mersmann. The fact and fiction of Martian dust storms. September 2015.

27 NASA. Mars report: Dust storms on Mars. https://mars.nasa.gov/resources/26555/mars-report-dust-storms-on-mars/#:~:text=January%202022%2C%20a%20dust%20storm,one%20another%20through%20the%20storm., 2022.

28 Robert P. Mueller, Tracie J. Prater, Monsi Roman, Jennifer E. Edmunson, Michael R. Fiske, and Peter Carrato. NASA centennial challenge: Three dimensional (3d) printed habitat, phase 3. pages 1–13, Washington, D.C., October 2019.

6: FOUNDATIONS FOR HUMANS ON MARS

Mark K. Moran, B.S.A.E., M.S.P.H.
Sterling, VA
breadwinproductions@gmail.com
Anjali Saini, Ed.S., M.Phil
Ashburn, VA
asaini@nvcc.edu

THE THREE MISSIONS

- Excavate about 100 foundations across 20 SWIM-Located Sites and shoot an ortho-mosaic aerial map of the results for each foundation)
- Seek out Past or Present Life (Astrobiology and GLASTR Lab)
- Wary of forward contamination, let's judiciously expose terrestrial life in controlled fashion to Mars regolith and atmosphere (SETBIO2MARS research arm)

EXCAVATIONS

In order to make inroads and pave the way for human exploration of Mars, robotics of this proposal will excavate into the Mars regolith and construct a series of architectural foundations, starting with shallow, small, and minimalist foundations which will incrementally step up to deeper, larger, and more sophisticated geometries whose surfaces are also encased/layered with in-situ derived cement or brick & mortar. Each site's elevation and local topography will help determine its place in this sequence, where generally higher elevations will precede generally lower elevation. A Mobile Carrier Vehicle will establish a Base Camp for a limited number of excavations at a particular site, while meanwhile a life science program unfolds. Upon completion of excavations and objectives at the first base camp, the Mobile Carrier will gather the drones together once again and migrate to a new Base Camp. There the process repeats, with incremental upgrades to the design of the construction foundations. Such foundations are deployed so that one or more future manned expeditions can pick and choose from among its fruits (100 foundations over the course of the nominal mission) to reduce the complexity of their own later missions.

Excavation Procedure

1. Follow the H_2O = follow the SWIM (Subsurface Water Ice Mapping on Mars) data base
2. At a series of sites from SWIM's best available water-ice locations, bring a platoon of robotic excavators in a mobile carrier vehicle. Drill heated bore holes.
3. At the earlier excavations, dig simple, shallow foundations - nothing fancy
4. At later excavations (Fig. 1), dig more deeply, add ISRU cement or brick & mortar (ie. get fancier)
5. Carry out ortho-mosaic aerial mapping of each dug site

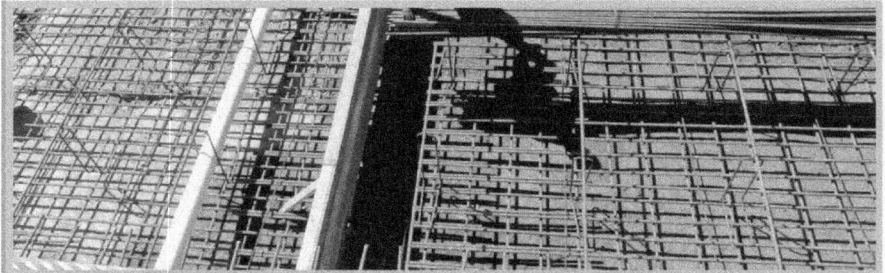

ISRU and Rebar in Foundations:
Cement can be a combination of Earth-transported ingredients mixed with ground Martian regolith

Fig 1 Later foundations will deploy In Situ Resource Utilization (ISRU)

One of the final, sophisticated foundations can be designed to abut the base of a scarp (Fig. 2) to reach a happy compromise between the safety advantages of underground shelter (spacious and highly protective against radiation exposure) and the construction freedoms of unconstrained, out-in-the-open flat terrain.

Fig. 2 Taking advantage of topography at base of scarp

Ortho-mosaic Aerial Mapping of Foundations

Following the excavation, any of our 8 RIADs can take imagery which the C³SEFAR can individually adjust and sew together into a composite map of the entire foundation (Fig 3). Do this before completion and after completion.

Fig 3 Mapping the excavation

The workhorses of the mission are 13 units of MGrODS (Fig. 4)

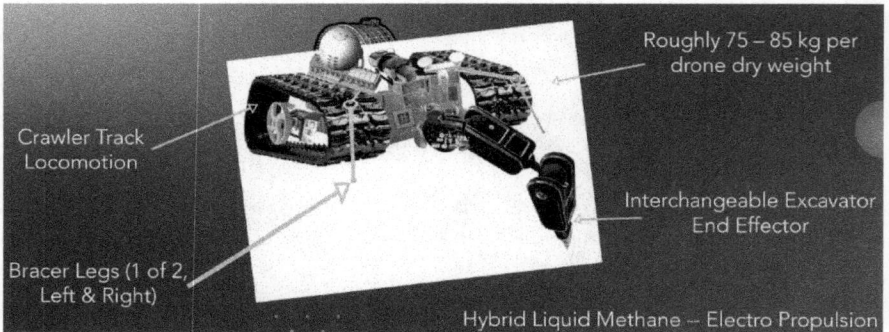

Roughly 75 – 85 kg per drone dry weight

Crawler Track Locomotion

Interchangeable Excavator End Effector

Bracer Legs (1 of 2, Left & Right)

Hybrid Liquid Methane – Electro Propulsion

Fig. 4 The main robotic components of the Midscale Ground Operations Drone (MGrOD)

EARTH BIOME EXPOSURES

Although missions to Mars are routinely sanitized, the paramount question is not "whether" our Earth-launched spacecraft and rover technology will eventually export, albeit unintentionally, Terrestrial biome life to Mars and contaminate it but *when* will such importation and contamination occur, if this has not already occurred? Given this jarring incongruity, better to conduct science that manages Terrestrial-exported organisms well in advance of a crisis and establish a clinical, foresighted approach than to blind our astrobiological investigations to everything except Mars indigenous life such as any that may be extant, and ignore the consequences once Terrestrial biome organisms stow away to Mars. This principle means that life research in situ must consist of not only an astrobiological arm but also a Terrestrial biotic arm, located as far distant from each other as is practicable.

Not only is the assumption foolhardy that no contamination from Earth biome sources will arise, but equally foolhardy is the assumption that when such contamination eventually occurs science will, without foresighted efforts like this, gracefully know how to handle the situation. Invasive contamination may already be at work on Mars from Earth technologies inadequately sterilized and, even if this has not yet taken place, is increasingly likely to occur in the near future based on successive, and larger scale, contact between our technologies and the surface of Mars [1]. Instead of blissful ignorance, the safer approach is to deliberately bring a limited scale, or miniature, contained biota or biome of Earth species micro-organisms (including some extremophiles and ammonia-oxidizing archaeon, but not including *Tersicoccus phoenicis* which grows in spacecraft clean rooms despite all efforts to sterilize [2], while in addition also bringing at least two or three small mammals) and methodically cultivate an Earth biota on Mars, while securely separating it from unsupervised contact with the Mars surface and instead introducing to it from the Mars environment such things as larger and larger exposures to regolith and atmosphere from that environment. If this Supervised Exposure of Terrestrial Biome to Mars Atmosphere Regolith and Setting

(SETBIO2MARS) laboratory can be secured against cross-contamination from indigenous life, then scientists can hope to learn from it how to accommodate the inevitable contamination of Mars from Earth biome species. If the system employs differential pressure concentric containment collars of security, the capability to count microorganisms within a given containment collar (enhancing detectability through radioactive carbon-14 in nutrient broths), and sufficient alarm sensors, a significant breach of the containment can trigger a mechanism for killing off the organisms if appropriate.

DETECTING PAST OR PRESENT MARS LIFE

- Carbon radioisotopes on Mars already show an isotope imbalance more in line with our experiences with life on Earth. Extend these efforts!
- Methane and oxygen gases have been detected in the atmosphere. Seasonal monitoring (SMOACH)!
- Gil Levin's experiments can be made more sophisticated, including control over chirality in nutrient broths
- We will need drills and spectroscopy lab run by our Carrier Vehicle to analyze the tailings

FORWARD CONTAMINATION RESEARCH

- Bring a number of different species of microbes in a thriveable culture. Divide into controls (not to be exposed) and treatments (to be exposed). Start with very small exposures to the treatment group and carefully monitor both groups.
- At a series of planned increments, raise exposure & track numbers & metabolism of microbes. Do quick, large dose exposure as precaution in case the mission were to be aborted. For the rest of the mission, do not rush exposure but slowly find out what happens. Tracking metabolites will be easier by use of Carbon 14.

WATER ICE AND ASTROBIOLOGY

This proposal advocates for following subsurface water ice, sampling this where possible by borehole drilling at SWIM-identified locations [3]. Near these locations, several science campaigns are carried out searching for signs of Mars indigenous life of the present or past. Among these investigations are carbon isotope ratio tracking, updated Labelled Release Experiments, and periodic tracking of atmospheric oxygen and methane.

BASE CAMPS

The mission proposed is (1 of 3) to leverage precursor investigations of Mars, (2 of 3) to conduct new investigations which coordinate across a suite of robotics and drone laboratories, and (3 of 3) realistically make inroads and pave the way on Mars for human exploration and eventual settlement. The central platform for airborne and ground robotics is a mobile 6-wheeled Carrier capable of moving these assets (with two exceptions) en masse from one location to another. In addition to vehicular assets (airborne drones and ground drones) are stationary drilling platforms which can be relocated as necessary using the same mobile carrier. The Mobile Carrier itself is moved only at intervals during an extended series of nominally 20 planned Base Camp migrations and occasional emergency operations. Once the Mobile Carrier has arrived at the next home base, it extends stabilizer jacks and settles as snugly as possible onto the surface with wheel chocks along the lines of earthly RV camping.

Migration of Base Camp
- 20 Such Migrations are planned: each along a SWIM route, taking account of topography and other constraints
- Early sites host excavations for shallow, uncomplicated foundations
- Advanced, later sites host deeper, more sophisticated foundations
- Safety precautions are emphasized so that excavated holes are not death traps for drones

The Mobile Carrier traverses the region of the landing site according to a series of steps. Initially, the Mobile Carrier is parked and anchored at the first of nominally twenty Base Camp sites within a favorable band from the Subsurface Water Ice Map (SWIM) [4]. At a varying radius on the order of between 50 m and 100 m, taking into account local topography, a perimeter around the Base Camp parked Carrier is established as a reference for reconnaissance, drilling sites for scientific purposes, drilling sites for boreholes heating or extracting water from subsurface water ice, and drilling sites to clarify the immediate location's excavatability. On this same perimeter is weather monitoring and the two bioscience facilities (see below: GLASTR laboratory at one azimuth and SETBIO2MARS laboratory at 180 degrees opposite azimuth, each of which deploys its own solar panels). A detachable greenhouse laboratory is housed within the SETBIO2MARS.

Astrobiology at Camp
The GLASTR laboratory will serve three primary objectives at each camp site: First, to catalogue carbon isotope proportions as motivated by a decade of observations by Curiosity Rover summarized in the 2022 Proceedings of the National Academy of Sciences article "Depleted carbon compositions observed at Gale Crater, Mars" by Christopher H. House, Gregory H. Wong, et al, January 25, 2022[5]; Second, to conduct an updated, improved Labelled Release Experiment (a la Vikings 1 & 2) [6] at each Base Camp; and, third, to track seasonal levels of atmospheric oxygen and methane.

The Mobile Carrier is also equipped with Solar Panels which can be cleaned periodically by (primarily) controlled electrostatic discharge or, otherwise, aimed downwash from Reconnaisance & Inspection Airborne Drones (RIAD). A general science laboratory, a spectroscopy laboratory, and an ISRU laboratory (includes studies of mortar and brick from in situ regolith and and using enzymes from Earth to liberate and collect oxygen from perchlorates) are located inside the carrier. A flood light from the Mobile Carrier is equipped to illuminate large swaths of within-perimeter areas.

Based on pre-launch analysis of the local terrain, the first of 20 Base Camps of the Carrier is designated. Sites should be chosen in part on the basis of either SWIM maps or Mars Odyssey's GRS detection of mid-latitude subsurface water ice [7]. From the designated first site, nominally five construction foundations, or attempted foundations, for a future manned settlement are excavated into the terrain, backhoeing/bull-dozing regolith in the process. This robotic task with state-of-the-art hybrid liquid methane engine technology will consume several times the amount of time as a human crewed, Earth-based effort.

SIZING OF EQUIPMENT

The standard human driven earth dozer/backhoe payload ranges from 3,000 to 5,000 kg [8]. The Mars MGrOD dozer/backhoe haul payload is 75 kg. Therefore, the number of MGrOD dozer haul payloads in on Earth dozer payload = 4000/75=54 (rounding up). About how much time does an Earth Payload excavation and unload/fill take for 20 m of travel? With caution, estimate 30-40 seconds. On Mars, our MGrOD doing 20 m of travel and its own 75 kg load will take two minutes for each such maneuver. If we assume the entire site would need 40 such maneuvers of an Earth dozer, then on Mars we will need 40 x 54 = 2,160 maneuvers of our workhorse drone which would each take 2 minutes apiece. Therefore, we will require 2,160 maneuvers over 4,320 minutes. This is 72 hours of robotic work. If there is downtime for cooling off, and we add ISRU cement or brick & mortar o the bottom of the foundation during one more hour and work is strictly during daylight, then this effort will require 1.5 weeks of work. 100 such foundations at 20 Base Camp sites will require roughly a full Martian year (two Earth years). Note: If at least 2 Mars backhoe/dozers work at the same time during excavation, this time can be cut in half or smaller.

Results may alter mission scope

If the effort to complete the first 3 or 4 pre-planned construction foundations fail, then fallback sites in the same general vicinity are resorted to. After the foundation or foundations have been laid for at least one successful or partly successful foundation (up to a maximum of 5 such foundations) and other mission objectives at the location met, then the entire Base Camp is uprooted (with the exception of the SETBIO2MARS, which is sustained at its original, permanent location and powered by its own dedicated solar panels), the Mobile Carrier packed for travel, and a new Base Camp at a different location sought out and secured. From the new site further foundations for a future manned settlement are again laid out, on different terrain, using a different design and scale (small scale in early excavations, larger at later ones) as well as different construction methods and ISRU, eventually laying on the bottom ISRU-derived bricks and/or concrete.[9] These migrations of the

Base Camp continue, with further foundations laid, slightly deeper, with further bricks or a concrete layer, until a minimum of 100 foundations are excavated during the course of one Martian year's lifetime mission. The motivation for the 100 foundations is to give future manned settlements a variety of options for exploiting this advance work and selecting the best site, or exploiting lessons learned, according to their capabilities, resources, needs, and circumstances.

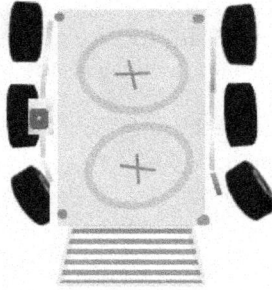

Fig.5 Top View of Mobile Carrier Vehicle

For each base camp, the three mission guidelines of leveraging past explorations, of conducting new investigations coordinating robotics and drone laboratories, and of breaking new ground for human exploration and eventual settlement of Mars orchestrate the entire program of effort. Two of the core leveragings of past astrobiology science investigations will be to extend further the efforts to track isotopes of carbon as possible indications of present or past life on Mars (as exemplified in the results from 9 years of samples from Curiosity's MSL, August 2012 - July 2021, in Proceedings of the National Academy of Sciences January 25, 2022) [10] and to track seasonal phenomena, especially atmospheric methane and oxygen. Both leveragings will be implemented on SWIM-targeted terrain.

Second, the robotics ensemble will be the foundation to conduct new investigations which coordinate across a suite of robotics and drone laboratories, including such efforts as 3D printing of laminated solar photovoltaic sheets, 3D printing of simple spares or exploiting new fiber-like parts which can flex in several ways, and the use of gelatinous resin and lasers to enable 3D prining without the requirement to build on a layer-by-layer only basis.

Third, to make inroads and pave the way for human exploration and eventual settlement of Mars, a program of architectural foundations, which start simple and shallow and gradually attain iterative sophistication and scale, cemented or bricked over using in situ resources for the sophisticated foundations, and increasing depth, along with drilling of boreholes by a heated boring rig to extract subsurface water, will serve this purpose.

The Mobile Carrier Vehicle is equipped with the *following telerobotic technology,* all of which are painted with bright IR and UV paint schemes and equipped with accelerometers and motion sensors:

- 1 high altitude reconnaissance proof of concept balloon drone (HARPOCOB)
- 8 reconnaissance & inspection airborne drones (RIADs squad) 4 reconnaissance & inspection ground drones (RIGDs squad), which are fastest- and farthest-ranging of the ground drones, with 4- or 6-legged locomotion (not wheeled)
- 13 midscale ground operations drones (MGrODs platoon), each more or less the size of a large microwave oven, with crawler tracks (some with wider and some with narrower treads) like a backhoe/bulldozer supplemented by 2-legged bracing/stabilizing legs which can also supplement tank-like locomotion or help free itself from getting stuck, a 6-dof boom arm with a load capacity/payload 70% of drone weight [drone weight is 75 kg, so 70% of this is 52.5 pounds] and a controllable lift radius, a sturdy backhoe, a heavy undercarriage, and rich AI capable of teaching such tasks as motion paths to lesser drones in record-playback mode. Ten of the 13 MGrODs have detachable dozer blades and detachable backhoe buckets. Twelve of the 13 MGrODs are flown within the hangar of the Mobile Carrier Vehicle inside the payload fairing, the 13th being flown in the payload fairing outside of the Mobile Carrier Vehicle.
- 5 tow troubleshoot & maintenance drones (TROMa squad)
  30 cell phone scale drones (CPSDs platoon), (designed on the basis of a separate Mars Society competition)
- 5 stationary but relocatable methane-powered drilling platforms (SRDPs) which can measure the excavatability of regolith, borehole drill into terrain with high subsurface water ice content and extract scientific or subsurface water ice samples.
- 4 relocatable stationary weather monitors with landline connection to Carrier (RSWMs), and which may sample atmosphere for SMOACH (see below).

TOTALS: 61 deployable drones (+ 4 weather monitors and 5 SDRPs), all powered by solid state batteries, apart from the Mobile Carrier itself, which mainly employs Fuel Cell power generation.of which 30 are mobile phone scale technology in 5 different "flavors" chosen by the Mars Society from a separate, dedicated Mars Society Competition (they crawl, walk, slither, etc by various forms of locomotion), 9 can fly (one of which is a proof of concept for a balloon drone), and the remaining 22 are the workhorse ground technology

TOTAL DRONES: 61, or 62 including the Mobile Carrier. One of which is a proof of concept (balloon technology).

LABORATORIES AND POWER SOURCES:

- Gil Levin Astrobiology or GLASTR LAB
- Updated Labelled Release Experiment (ULREX)
- Isotopes of Carbon Tracking (ISOCART)
- Seasonal Monitoring of O2 and CH4 (SMOACH)
- Supervised Exposure Of Terrestrial Biome To Mars Atmosphere Regolith & Setting Lab or SETBIO2MARS Lab (within which initially is housed the Detachable Greenhouse Lab or DEGRL) equipped with wireless telemetry and negative pressure containment collars and its own independent power supply and photovoltaic solar panels
- Carrier Engineering & General Science Lab or CAREGENS Laboratory
- Spectroscopy Lab or SPECL ("speckle")
- Seismology Lab or SEISL ("sizzle")
- ISRU Lab for grinding regolith and mixing with liquid to produce Mars bricks and Mars cement analogue out of basaltic fiber from regolith, second, (proof of concept) laminar solar photovoltaic cells for solar panels, and third enzymes used to liberate and store oxygen from perchlorates
- LDEF-Inspired or Diverse Materials Prolonged Exposure To Mars Sample Return or DMPEXMSR (i.e., Long Duration Exposure Facility - Modelled Leave Behind For Sample Return)
- Carrier Vehicle Methanol Fuel Cell Power Supply (onboard the Carrier) [Why not Radioisotope power supply i.e., RTG? For costs, see the document: https://inldigitallibrary.inl.gov/sites/sti/sti/7267852.pdf][9]

Table 1 Comparative Estimated Costs
Base Camp Power Plant utilizing RTG vs Base Camp Power Plant utilizing Fuel Cell

RTG = Main Power Plant for Base Camp   Fuel Cell = Main Power Plant for Base Camp

RTG	$ 69 M	Fuel Cell	$ 5 M
WORKHORSE DRONES	$ 18 M	WORKHORSE DRONES	$ 18 M
CPSDs	$ 2 M	CPSDs	$ 2 M
MOBILE CARRIER LESS MAIN POWER BUT INCLUDING SOLAR BACKUP	$ 30 M	MOBILE CARRIER LESS MAIN POWER BUT INCLUDING SOLAR BACKUP	$ 30 M
STATIONARY HARDWARE	$ 5 M	STATIONARY HARDWARE	$ 5 M
OTHER ASSETS (such as GSE on Earth)	$ 2 M	OTHER ASSETS (such as GSE on]Earth)	$ 2 M
TOTALS:	$ 126 M		$ 62 M

$126 M / $62 M = 2.03 i.e., the RTG Main Power Source would double the total cost of the program in its entirety compared to using instead Fuel Cell Technology. Therefore, it is recommended that the Fuel Cell be utilized, with solar panel backup power supply, rather than expensive RTG technology with solar panel backup power.

Command Control Comms and Sensor Farm or C³SEFAR

An artificial neural net or ANN-unit modular, parallel processing AI suite is overseen by a programmable supervisor. Intel Labs currently offers a second generation neuromorphic chip called Loihi 2 able to be adapted to this function. The C³SEFAR communicates directly via steerable antenna to TDRS/Earth or indirectly via Mars Orbiting Satellite, controls and operates the Carrier, and coordinates the 22 work-horse ground technology, the 8 flying RIADs, and the 30 CPSDs, either directly or via the mediation of either the 13-unit MGrODs or the 5-unit TROMa Squad. C³SEFAR is equipped with an advanced LIDAR.

Telecom between the Carrier's C³SEFAR and drones consists of three modes of bandwidth: 1.) Narrowest bandwidth dedicated to routine checklist reporting of status, location, etc., 2.) Mid-level bandwidth for real-time remote instructing and feedback loop reporting of operations controlled by the C³SEFAR or troubleshoot report, and 3.) Widest bandwidth telecom for upload/download of data intense transmissions (e.g., video imagery). Telecom between the Carrier's C³SEFAR and Mars Satellites consists of NASA standard modes of bandwidth. Direct-to-Earth telerobotic operations, as opposed to primarily autonomous or semi-autonomous operations, are nominally limited to migrations of the Mobile Carrier Vehicle, early trial runs of MGrOD excavations, and emergency operations that are not time-urgent.

The C³SEFAR is mounted on the upper right hand side from the Carrier's point of view, amidships. Of the 13 MGrODs, 6 are equipped with portable drills and abrasion devices to gather specimens from surfaces encountered at random locations at a distance from the perimeter, either to add to the SETBIO2MARS exposure stockpile or for other laboratory use. For sample collection, the MGrOD must always tag the sample with its stratigraphic context. These drills are capable of grading the difficulty of excavation at the location. The other 7 MGrODs have the capability to transport out or retrieve up to 10 CPSDs in one trip. Ten of the MGrODs have a front-end, very wide bull-dozing blade with wings to prevent spillage at either side (payload of 40-52.5 kg) as well as left-hand and right-hand bracer legs (mountable toward either front or back of crawler tracks).

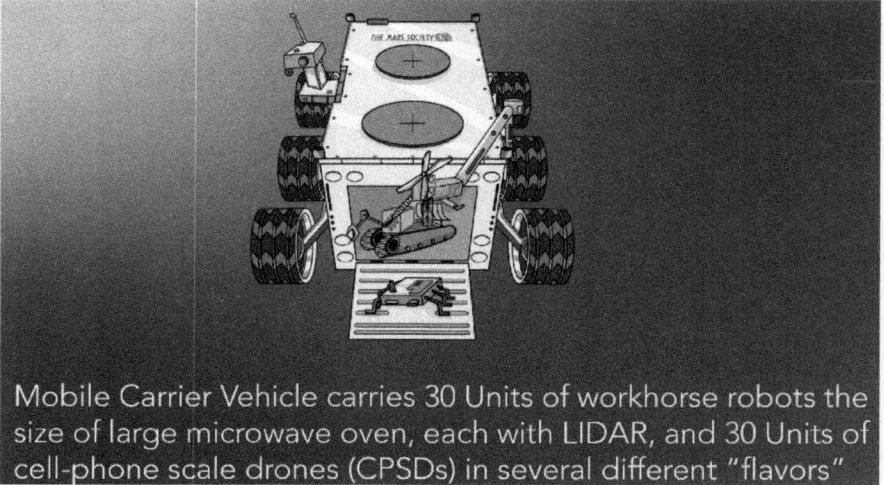

Mobile Carrier Vehicle carries 30 Units of workhorse robots the size of large microwave oven, each with LIDAR, and 30 Units of cell-phone scale drones (CPSDs) in several different "flavors"

Fig. 6 Carrier with hanger door open revealing an MGrD (on mouth of hangar door and a CPSD on ramp

The MGrODs are crawler track propelled, left and right sides, plus two legs for support and bracing while excavating. MGrOD has a sturdy backhoe. The MGrOD is *hybrid internal combustion (methanol fuel and oxidizer equipped) as well as electric equipped* with battery charging and wired (touching) telecom port (2-way in case of assisting another drone) as well as cameras. The MGrOD is the size of a bulky microwave oven. Each MGrOD is equipped with LIDAR, at least one 6-dof arm, dozer blade, front-end loader or backhoe with a payload 70% of drone weight (52.5 kg), and rich AI capable of teaching such tasks as motion paths to lesser drones which operate in record-playback mode while being taught. MGrODs are equipped with a small overhead solar panel which has a limited ability to recharge its batteries or be used in an emergency. MGrODs are equipped with LIDAR and a blade-mounted (or hoe-mounted) sensor and laser level.

The TROMa's are each large enough to transport one entire midscale-size drone or 20 CPSDs during one trip. Their travel range is longer than all drones except for the RIGD drones, which have the same range. This range should be at least 20 times the Perimeter mean radius. TROMa's can exchange a drone's rechargeable batteries with their onboard cache of rechargeable batteries or park nearby the other drone and recharge the other without removing batteries. They are also capable of in situ/on-site maintenance of the locomotion systems for any drone. They can plug into and troubleshoot computer hardware and software issues. TROMa's are equipped with a solar panel twice the size of an MGrOD's which can recharge their batteries and be used in an emergency. TROMa's also include LIDAR. A TROMa can also bring fuel or oxidizer for an MGrOD or for a SRDP.

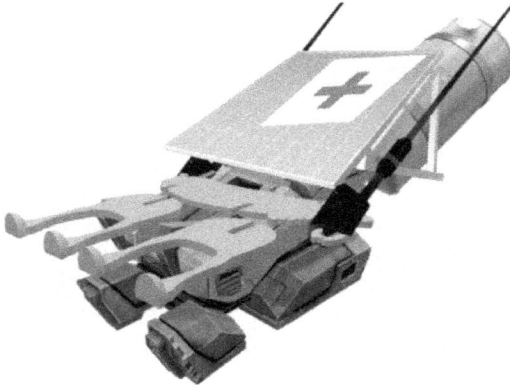

Figure 7 TrOMa drones perform trouble-shooting, maintenance, and drone transport

The 8 RIADs and 4 RIGDs are capable of exploring the furthest distances from the Carrier Base Camp and traveling at the fastest rate of all the drone technologies, with LIDAR. These reconnaissance drones have a number of autonomous features to be used to reconnoiter and survey as needed but are mainly controlled in real time from the Mobile Carrier and make use of radio beacons to help determine their locations, both RIADs and RIGDs. Nominally, RIGDs can travel up to 20 times the Perimeter's mean radius. RIADs and RIGDs are equipped with solar panels which can recharge their batteries and be used in an emergency. Landing legs (4) of RIADs are spread wide, so that in the middle of each leg's full length the pole is not overly steep in angle and attached to the top of each will be one of four narrow and long, aerodynamically contoured solar panels at this modest angle, angled or curved based on Earth testing including ground effect testing. Both RIADs and RIGDs have every possible communications enhancements and equipment. The primary mission for a RIAD encompasses six roles: 1.) the support of routine telecom, 2.) restoration of interrupted telecom between Mobile Carrier C^{3}SEFAR and other drone(s), 3.) emergency telecom services, 4.) reconnoitering terrain that has not yet been inspected closely or that lies ahead either of the Mobile Carrier's path or of an important drone's path including excavation, with surveillance videography or photography, 5.) inspection from above toward any drone, landscape feature, or other object of interest including excavations, with videography or photography, and 6.) using downwash to clean off solar panels (as a secondary way to do this after electrostatic discharge cleaning).

As a rule, MGrODs are required to travel within 40 m of a "buddy" MGrOD. The equivalent rule for CPSDs is that at all times at least 4 CPSDs should be within a radius of 5 m from one other.

The 30 CPSDs, in 5 different designs or "flavors," each of them in several different configurations, are intended to be deployed in platoons or squads of between 6 and 12 units, which swarm around a particular Site of Designated-Emphasis or SoDE.

During construction foundation work, CPSDs can rove around service locations slated to be dug up during the foundation dig, scouting for large rocks or other hindrances for the equipment, and in certain cases actually moving gravel. CPSDs are NOT equipped with a solar panel.

30 CPSDs in 5 "Flavors":
- 8 of 1st Flavor (Surface Remediation and Hangar Cleaning Duties)
- 5 of 4" by 6" Origami A AKA 2nd flavor (Foundation Support Operations) capable of folds into 10 shapes for 10 distinct operational functions, including planting of safety markers
- 5 of 3" by 5" Origami B AKA 3rd flavor (General Purpose and Fuel Cell Support) capable of folds into 10 shapes for 10 distinct operational functions
- 5 of 4th flavor (Tight Space Navigating, Retrieving or Depositing, and Within & Nearby Carrier Ops including Electrical Duty)
- 7 of 5th flavor (Periphery Operations and Astrobiology Support Operations)

On the Carrier, at first it seemed that there should be an elevator to bring RIADs between the landing pads (2 of which sit on the roof of the Mobile Carrier) and the hangar interior to the Mobile Carrier. However, because of dust incursions potentially playing havoc within the Mobile Carrier, it has been decided that the transport of RIADs into and out of the hangar to the landing pads instead should be handled by two crane arms mounted on the outside of the Carrier. See sketch of Carrier lifting a RIAD.

The Hangar within the Carrier is equipped with Bridge Gantry Crane technology, a Gantry type Crane hoisting from the ceiling that can lift and lower relatively heavy hardware, including RIADs, RIGDs, MGrODs, TROMas, as well as much lighter CPSDs from one pair of X, Y coordinates to another along the floor. There is an overhead flood light inside the Hangar, a pointable flood light in the forward direction of the Mobile Carrier, as well as cameras.

The fully deployed Base Camp is the Carrier-Centered region consisting of the Carrier itself, a perimeter at least 50 m from the Carrier and at most 100 m from the Carrier, at least two positioned RIADs or RIGDs on the periphery, the SETBIO2MARS (with or without the DEGRL), the GLASTR stationed 180 degrees apart from the SETBIO2MARS on the perimeter, and a clearly marked with lights Backup Improvised Landing Site about 6 or 8 m from the Carrier available should the two rooftop Carrier pads be temporarily unavailable, any drill sites especially for excavation of a new foundation, warning lights and signals along key points of an excavated foundation, and the entire area within the perimeter. The entire Base Camp covers an area between about 30,000 square meters and 126,000 square meters. Exact boundaries of the perimeter of the camp depend on the terrain as found, including obstacles, shifts in elevation, and other unique features.

<u>3D Printer</u>
Source code is available to the public for the following highly versatile 3D printing scenarios which are beautifully applicable to robotic technology deployed on Mars. https://youtu.be/gfMyGad1Gmc

Operational Rules
Except for TROMa's, any drone in operation must at all times maintain position within 40 m of a "buddy" drone of the same design or better. If the drone is a CPSD, then it must at all times be within 10 m of at least 3 other CPSDs (or under emergency conditions at least 2 other CPSDs). Regardless of mission, a platoon of CPSDs must at all times operate within the 40 m range of two of the larger drones each within their buddy range of the other. At all times drones must maintain direct telematic contact with the Mobile Carrier, or facing a hardware failure must relay important data via the largest locally available buddy drone, which in turn must cooperate. The Mobile Carrier Vehicle itself must relay information either directly to Earth control (when possible at times) or indirectly to a Mars orbiting platform within communication range (this is also only available intermittently). When the sun is in the line of sight to the Earth, direct communications are impossible. During such interruption, all operations are autonomous, semi-autonomous (supervised by the Mobile Carrier's C^3SEFAR), or suspended. All drones must be equipped with safeguards against blinding of sensor instruments.

Spot lights from drones must be pointed strictly toward the work site or a SoDE. Solar panels shall be shielded and grounded to reduce single event upset events (SEUs) and non-cleaning electrostatic discharge.

One MGrOD may be deployed for a particular operation and then returned to Carrier and replaced by a tag team other drone.

The above operations can be combined, for certain scenarios, although certain combinations are categorically impossible (e.g., parked in Carrier and beyond-perimeter site ops).

Single Event Upsets (SEUs) happen because of high energy charged particle cosmic events (calling for EM Shielding and Grounding). The Mobile Carrier and all MGrODs must be shown to operate in an uninterrupted fashion without system-level failure from the same number of SEUs as at least the Perseverance Rover was designed for.

Electrostatic Discharge is a risk faced, most especially, by solar panels.
EMI engineering shall be carried out to reduce unintended antenna effects and garbling of instructions. At the same time, planned electrostatic discharge will be the primary mode for cleaning dust accumulation from the Carrier's solar panels, backed up by RIAD downwash.

Servicing (such as direct contact re-charging or changing a mechanical joint or locomotion part) can release particulates into the environment. Such particulate releases shall be marked on the Mobile Carrier Vehicle's event log.

Fiber Optic Data buses have the advantage that they do not need grounding and should be used where possible.

C³SEFAR Conceptual Requirements
- Must track the entire inventory of all drones, labs, resources, and items whether installed, stowed in the hangar, or undergoing maintenance/repair/servicing, using radiobeacons, special reflective paints and corresponding EM frequencies, visible contact, and radio telemetry
- Remotely operate any work-horse/backhoe-dozer MGrOD operating within the periphery of the Base Camp, at least during trial run excavations at earlier base camps, allowing for detail level operations to be handled by the drone itself, and trial run algorithms be maintained and fine-tuned with telerobotic Earth instructions for independent run tests
- Drones operating outside the perimeter must have a high degree of autonomous control
- C³SEFAR is the primary contact between the entire Base Camp including all drones and either Earth contact or Mars satellite contact, although drone-to-drone direct communications shall be available when needed
- Fiber optics will be as much as feasible the primary internal data transmission bus
- Remote testing and operation of the HARPOCOB will be a function of the C³SEFAR, including the ability to continue proof of concept testing thereof or to abort the proof of concept
- C³SEFAR shall have a backup auxiliary control system to be used temporarily while resolving any C³SEFAR operational snags

3D Printer List of Jobs
- Organic LED (OLED) printing of lighting devices[10]
- printing of Solar Photovoltaic gallium arsenide Sheets
- printing simple spare parts and Tool Effectors
- printing of ISRU Bricks for Foundations
- flexible items that involve flexible operations (see 3D Printer, page 7)
- using gelatinous resin & lasers, can 3D print point to point and not be limited to layer by layer

Supervised Exposure Of Terrestrial Biome To Mars Atmosphere Regolith & Setting or SETBIO2MARS

The SETBIO2MARS Lab consists of facilities to research a limited number of microorganisms, a limited number of plants, and one mammal species imported from Earth, exposing them to Mars substances and conditions without release into the wild, and the Detachable Greenhouse Lab or DEGRL. The DEGRL is contained within the SETBIO2MARS because it too deals with Terrestrial Biome organisms albeit of plants that are larger. The DEGRL is detachable so that, in the event the Base Camp relocates, the DEGRL can be brought with it or, optionally, left in place. The SETBIO2MARS as a whole is never moved once its location is established. For a starter list of species to be included in the SETBIO2MARS, see table.

Table 2
SETBIO2MARS Earth Biome Species

Mammal (Eukaryote)	wildtype mice (2 male and 2 female)	During interplanetary transport, such mice should be kept in a state of synthetic torpor
Bacterium	Shewanella oneidensis	
Bacterium	Paramecium	
Fungus	Cryptococcus neoformans	Found at Chernobyl. Thrives in gamma radiation.
Archaeon	Thermococcus bergensis	
Archaeon	Haloarcula sebkhae	
Archaeon	ammonia-oxidizing archaeon	
Plant (for DEGRL)	bean sprouts	
Plant (for DEGRL)	duckwood (Lemna minor)	
Plant (for DEGRL)	water fern (Azolla filiculoides)	
Lichen	Pleurosticta acetabulum	
WARNING: The species *Tersicoccus phoenicis* has been found in clean rooms and so is extraordinarily difficult to sterilize, even for advanced technology programs. It should *not* be included in the manifest.		

Gil Levin Astrobiology or GLASTR Lab

1. Updated Labelled Release Experiment (ULREX)

The Labelled Release experiment, deployed on Vikings I and II, scooped regolith samples, mixed them with a nourishment broth spiked with radio-labelled ^{14}C, and tracked the radioactive gases that were subsequently produced [11]. Its results were suggestive of life. However, skeptics pointed to perchlorates, or perchlorates irradiated over a long term, as an alternative explanation. However, analysis of the results included evidence of circadian rhythm as well as very low temperature thresholds that stopped the outgassing, both unlikely outcomes from non-living chemistry. In the ULREX improved version of the Viking Labelled Release experiment, the nutrient broths will come in left-hand chirality (amino acids), right-hand chirality (carbohydrates), and mixed chirality (mixture of the same) batches. ULREX will also repeat the experiment at a series of temperatures so that the effects of temperature are carefully tracked. In addition, a series of controls will be run through the same steps.

2. Isotopes of Carbon Tracking (ISOCART)

The RSWMs handle Seasonal Monitoring of O2 and CH4 (SMOACH) whose samples are sent to the SPECL lab for isotope processing and logging. Sampling of atmosphere can be taken from one of the Weather Monitors.

Table 3 Drilling Tasks for SRDPs, RIGDs, AND MGrODs

DEVICE	TASK
SRDP and/or RIGD	DETERMINE HARDNESS/ESCAVATABILITY OF REGOLITH
SRDP	BELOW GRADE SAMPLING
SRDP	DEEP SAMPLING
SRDP	PERIPHERY EARLY DRILLING
RIGD	BELOW GRADE SAMPLING
RIGD	PERIPHERY EARLY DRILLING
MGrOD	DETERMINE HARDNESS/ESCAVATABILITY OF REGOLITH
MGrOD	BELOW GRADE SAMPLING
MGrOD	MULTIPLE PROBE HARDNESSES
MGrOD	ROCK REMOVAL BY INSERT

Table 4 Tasks For Particular Drone Types

Scout out site for Base Camp	RIAD and RIGD (also optional, if available, is HARPOCOB)
Scout out site for Construction Foundation	RIAD and RIGD

Drill Hole into Rock for Science Purposes	MGrOD and SRDP
Harvest the Tailings from Drilled Hole & Collect in Labelled Tedlar (or Equivalent) Specimen Bag	MGrOD or (rarely) CPSD
Harvest Gas Immediately from Drilled Hole	MGrOD or CPSD
Harvest Atmospheric Gas Sample	RSWM
Secure Carrier to Surface for Base Camp Mode of Operations	MGrOD or RIGD
Keep Track of all Drones (including Status, Location, Plot, Realtime Objective, Other Objectives)	Carrier C³SEFAR
Check and Warn about Ground Path in Front of Other Drones	RIAD and RIGD
Locate and Assist Compromised Drone	TROMa
Return Compromised Drone to Carrier	TROMa
Deploy to Drill Site an SRDP	MGrOD or RIGD
Lift Disabled Drone from underneath	TROMa

SEQUENTIAL MIGRATION OF BASE CAMP TO BASE CAMP (SMBC2)

At the heart of top-level mission planning will be a map that takes into account the topography, important features of the terrain, and the current science campaign. For this proposal, the most important feature of the terrain in determining the course of waypoints of the base camp's sequential locations is the likely subsurface water ice content and depths. This information is most readily available by means of the Subsurface Water Ice Mapping (acronym SWIM) project. Access to the SWIM maps are high priority not only for the landing site selection but also for SMBC2.

In order to establish a precedent for SMBC2, below is an example map of a terrain from Valles Marineris canyon. The first base camp and the landing site should be as close to each other as possible. Following lanes interior to the SWIM identified lobes rich with subsurface ice at moderate or shallow depths, the objective of understanding how best to excavate foundations can be served by marking two or three "control" excavations on constant elevation contours, unburdened with the deployment of cement or brick and mortar, followed by a series of excavation sites that experiment with add-on sophistication, bands of elevation, or types of terrain. If each base camp will supervise a small cluster of these excavations, then the excavation sites within such a cluster must be within reach of the same base camp. The Valles Marineris sketch shown depicts an example of SMBC2 mapping.

The navigation plot for SMBC2 is based primarily on the SWIM maps for subsurface water ice. Within a broad swath for this path, the location of the targeted next Base Camp site in the migration sequence is fine-tuned based on 1.) terrain and suitability to foundation digs and borehole drilling, 2.) friendliness of the somewhat more distant surrounding terrain to movement and planned activities of drones of either size, 3.) avoidance of natural obstacles to nominal telecommunications, 4.) probability that the region is relatively friendly to any indigenous biome should such a biome exist or have ever existed, and 5.) minimizing inter-camp distance of travel for the Carrier (between preceding and following Base Camp sites but also taking into account the possibility that the next site in sequence will be rejected and an alternative site have to be selected). The migration itself is telerobotically controlled from Earth except during telecommunications blackout.

The migration of the Base Camp to the next in sequence site unfolds as follows:

1.) The Carrier Vehicle in motion is to stay within a lane highlighted upon the most current SWIM-provided map.
2.) At a line-of-site distance of 25-150 meters, for intervals during travel one RIAD flies ahead of the Carrier Vehicle to reconnoiter and warn of any looming issues.
3.) Although for short periods the Carrier Vehicle may, if necessary, advance without a RIAD preceding it, during migration most if not all of the time when the terrain is uncertain or presents dangers a RIAD shall nominally be flown ahead of it. When the RIAD's batteries are verging close to depleted, the Carrier Vehicle is to halt, the RIAD return and be brought inside the hangar for recharge, and a fully charged fresh RIAD is launched to continue as before.
4.) Upon reaching within a few hundred meters of the targeted next site, two RIGDs are to be deployed in addition to the RIAD. In order to consider the proposed location in greater depth, both the ground and airborne drones must reconnoiter sufficiently and confirm the suitability of (or fail to veto) the proposed site.
5.) When the reconnoitering has been unable to veto the site, more drones are deployed, and a stationary borehole drill (SRDP) is secured to a spot as close as possible to a first excavation of a foundation. Either this boring will serve as final confirmation of the chosen site, or if its findings are ambiguous, a second borehole will be drilled at a second spot.
6.) When reconnoiters and borehole both confirm the finalized choice of site, then the Mobile Carrier is parked at the most level available spot. Chocks are deployed at the Mobile Carrier's wheels, crude leveling is done, the internal Carrier levelling mechanism then implements finer leveling, and stabilizing jacks are extended and secured. A few CPSDs are deployed and operated primarily to clear out rocks of sufficient size from the immediate vicinity of the Carrier. At this point, the Carrier's solar panels are deployed. These solar panels are not the primary source of power, but a supplementary/backup source behind the Carrier's methanol fuel cells.
7.) Should the proposed site be vetoed, then all drones apart from one RIAD are returned to the hangar and revert to Step #1.

Otherwise,

8.) The next research campaign is commenced.

9.) When the research campaign has run its course, primary among the safety protocols shall be to safeguard against any drone before departure or the Carrier Vehicle upon departure venturing dangerously close to the excavations, either completed excavations or even aborted excavations.

10.) Until implementing the last of all Base Camps, return to Step 1 above, under telerobotic Earth control.

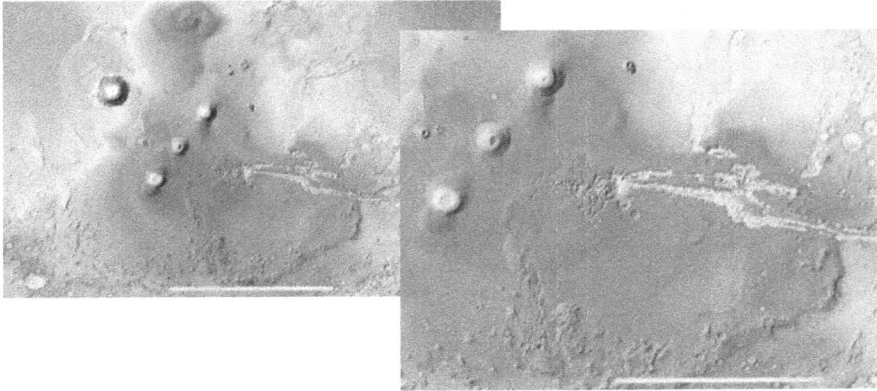

Fig. 8 Olympus Mons, Tharsis Ridge to Left of Valles Marineris.
Closer zoom in right hand figure.

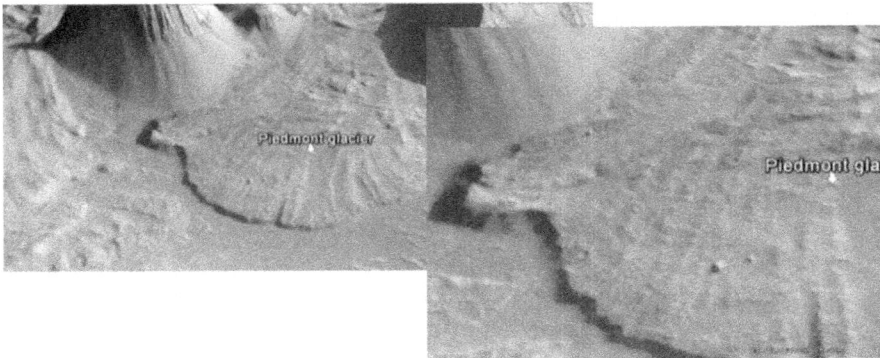

Fig. 9 Local Terrain At Canyon Floor, Wider And Zoomed In.

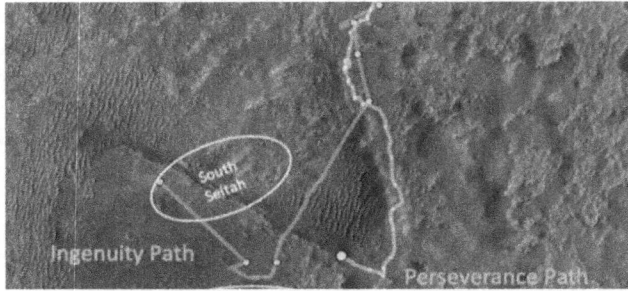

Figure 10. Example of how Perseverance and its Ingenuity drone were used in coordinated travel

DISCUSSION AND CONCLUDING REMARKS

This proposal's paramount mission: (1) to leverage precursor Mars expeditions, (2) to conduct new investigations which coordinate across a suite of robotics and drone laboratories, and (3) realistically make inroads and pave the way on Mars for human exploration and eventual settlement. Mission Guideline 1 is realized through new campaigns for life science investigations including an updated Labelled Release experiment, systematic introduction of Mars environment to Earth biome under controlled conditions, measuring isotopes of carbon, and seasonal observations of the Mars atmosphere. Mission Guideline 2 is realized through the use of new robotic drone and robotic laboratory technology of diverse configurations, advancing ISRU by making Mars bricks and mortar and liberating oxygen from perchlorates; searching for life by means of an updated labelled release experiment; and proof of concept testing of a balloon in the Mars atmosphere. Mission Guideline 3 is realized through an experiment laying down architectural foundations for structural habitats or facilities in the Mars regolith on an incrementally improving basis starting with small, simple, and unsophisticated excavations, and the practical introduction of a 3D Printer to replace classes of worn, fraying, broken, or damaged parts as well as proof of concepts for flexible solar panels. It is recommended that the smaller class of drone, the CPSDs, be designed by contestants competing in another competition sponsored by the Mars Society.

Although the excavation of foundations presents many unknowns and therefore risks, this systematic effort to lay down these foundations offers a profound opportunity for Mars exploration to advance beyond the robotics-only phase to a permanent manned habitation of the Red Planet.

Author of this proposal: Mark Moran, M.S. (B.S. Aerospace Engineering, Virginia Tech, and M.S. Biostatistics, University of South Carolina)

Holder of a patent pending racquet sport. Resident with family in Northern Virginia.

breadwinproductions@gmail.com

REFERENCES – Scientific, Engineering, And Mars Relevant Articles

1. https://cosmosmagazine.com/space/strict-rules-around-contamination-hamper-exploration-for-life-beyond-earth/?amp=1

2. See Vaishampayan, Parag et al. "Description of Tersicoccus phoenicis gen. nov., sp. nov. isolated from spacecraft assembly clean room environments." *International journal of systematic and evolutionary microbiology* vol. 63,Pt 7 (2013): 2463-2471. doi:10.1099/ijs.0.047134-0

3. https://swim.psi.edu/

4. See 3.

5. https://pennstate.pure.elsevier.com/en/publications/depleted-carbon-isotope-compositions-observed-at-gale-crater-mars

6. https://www.science.org/doi/10.1126/science.194.4271.1322

7. https://www.nature.com/articles/news020225-15

8. https://assets.cnhindustrial.com/nhce/APAC_Downloads/Equipment/Wheel-Loaders/W110C-W130C_ANZ3401NCGB.pdf

9. https://inldigitallibrary.inl.gov/sites/sti/sti/7267852.pdf

10. https://youtu.be/g3YAZZ486qk https://youtu.be/g3YAZZ486qk

11. https://www.pnas.org/doi/10.1073/pnas.2115651119

12. https://www.ncbi.nlm.nih.gov/pmc/articles/PMC6445182/

13. Science paper re clean room microbe
https://pubmed.ncbi.nlm.nih.gov/23223813/

14. Resistant microbe in clean rooms
https://mars.nasa.gov/news/1532/rare-new-microbe-found-in-two-distant-clean-rooms/#:~:text=Tersi%20is%20from%20Latin%20for%20clean%2C%20like%20the,test-swabbing%20the%20floor%20in%20the%20Florida%20clean%20room.

15. Extremophiles
https://www.newscientist.com/article/2306352-bacteria-survive-extremes-that-may-have-existed-in-ancient-mars-lakes/

16. Water ice buried on mars
https://doi.org/10.1093/acrefore/9780190647926.013.239

17. ESA document about margins:
https://sci.esa.int/documents/34375/36249/1567260131067-Margin_philosophy_for_science_assessment_studies_1.3.pdf

18. Borehole drilling and rehabilitation under field conditions
https://www.icrc.org/en/doc/assets/files/other/icrc_002_0998.pd

19. ISRU & 3D printing
https://www.sciencedirect.com/science/article/pii/S2666539521001620

www.ingramcontent.com/pod-product-compliance
Lightning Source LLC
Chambersburg PA
CBHW031857200326
41597CB00012B/458